Willi Darr

Concepts and paradigms in operative, strategic and social time management

Willi Darr

Concepts and paradigms in operative, strategic and social time management

tredition Verlag
Hamburg

Prof. Dr. Willi Darr
University of Applied Sciences Hof
Procurement and Logistics Management
Alfons-Goppel-Platz 1 ı D-95028 Hof/Saale

Hardcover   ISBN 978-3-347-03014-5
Paperback   ISBN 978-3-347-03013-8
E-Book   ISBN 978-3-347-03015-2

Bibliografische Information der Deutschen Nationalbibliothek
Bibliographic Information of the German National Library

Die Deutsche Nationalbibliothek verzeichnet diese Publikation in der Deutschen Nationalbibliografie; detaillierte bibliografische Daten sind im Internet über http://dnb.dnb.de abrufbar.
The Deutsche Nationalbibliothek lists this publication in the Deutsche Nationalbibliografie; detailed bibliographic data can be found on the Internet at http://dnb.dnb.de.

Cover and design by tredition & Willi Darr

Translated by Sally Rendl

Originaltitel: Konzepte und Paradigmen im operativen, strategischen und gesellschaftlichen Zeitmanagement

Herstellung und Verlag ı Printed and published by
tredition GmbH, Halenreie 40-44, D-22359 Hamburg, Germany

Contents

List of figures

List of tables

List of abbreviations

A.D., AD	Anno Domini
AtO	Assemble to Order
Aufl.	Auflage, Volume
B.C., BC	Before Christ
BtO	Build to Order
BtS	Build to Stock
BVerfG	Bundesverfassungsgericht
d	Dauer, duration
DVD	Digital Video Disc bzw. Digital Versatile Disc
E	Entscheidung, Decision
EDIFACT	United Nations Electronic Data Interchange for Administration, Commerce and Transport
EinhZeitG	Einheiten- und Zeitgesetz
EU	Europäische Union
Fn.	Fußnote, Footnote
GB	Großbritannien, Great Britain
GG	Grundgesetz
GMT	Greenwich Mean Time
GPS	Global Positioning System
GTIN	Global Trade Item Number
H.	Heft, Issue
i.e.	id est, that is
ILIPT	Intelligent Logistics for Innovative Product Technologies
INCOTERMS	International Commercial Terms
Iss.	Issue, Ausgabe
Jg.	Jahrgang, Volume
IoT	Internet of Things
K	Kosten (Prozesskosten), Cost
No.	Number, Nummer
OPP	Order Penetration Point
p.	Page, Seite
Rn.	Randnummer
S.	Seite
SDD	Same Day Delivery
t	Time
TAI	Internationale Atomzeit
TPS	Toyota Produktions System
UK	United Kingdom
US	United States
USP	Unique Selling Proposition
UT	Universal Time
UTC	Coordinated Universal Time
Vol.	Volume, Jahrgang
Z	Zeit (Prozesszeit), Time (Process Time)
ZeitG	Zeitgesetz

Introductory thoughts instead of a foreword

The subject of this book is a subject-specific business discussion on the paradigms of time management. The time itself is not tradable, not storable and also not purchasable to acquire. Time cannot be heard, smelled, tasted or felt. Nevertheless, it is omnipresent for all persons in companies as well as in private life. Some selected examples may prove this, and each reader may make its personal evaluation, to what extent the time and/or the time pressure releases an oppressive or relaxing feeling for it. Here are the examples from **daily life**:

- The alarm clock and the church bells ring at 6 o'clock in the morning.

- The children should get up on time to be at school or to reach the school bus.

- Employees should arrive early at the workplace. The schedule is well filled. This also applies to students: The lecture begins punctually at 8 o'clock.

- The morning traffic torments through the city and makes promised appointments in the early morning uncertain. The duration of the green phase at the traffic lights is very short and only a few cars manage to get further.

- The annual holiday period is several weeks and for the planning of the annual summer vacation, all colleagues have to meet a common regulation/ have to make a joint arrangement in coordination with their supervisor/ company.

- Today there is little time for lunch or lunch break. The next appointment is already "on the doorstep".

These examples make it clear that time plays a permanent role in everyday life. Possible time buffers can alleviate the temporal tension and defuse the time pressure (and vice versa). This is expressed in idioms or proverbs, such as having no time or losing no time or the race against time.

Time also plays a prominent role in **sport**. Not only does a football match last 90 minutes (plus injury time), but the fame and honour of the best competitive athletes in motor sports, skiing or athletics sometimes only depend on a few hundredths of a second. The tenth of a second is often no longer the yardstick to distinguish between victory and defeat.

Time also plays a prominent role in **art** and **literature**. The picture "La persistencia da la memoria" (the permanence of memory or the elapsing time) by Salvadore Dali from 1931, pictures on still life, the books by Michael Ende (Momo) and Thomas Mann (Der Zauberberg) or the films "In Time" or "Modern Times" with Charlie Chaplin prove this. The books on time travel with the time machines of H.G. Wells or on the circumnavigation of the world in (then) record time by Jules Verne ("Around the world in 80 days") are also world-famous.

Time is also currently being discussed: (i) Thus the time discussion in the state of Bavaria was again decided in school politics to the G8 or G9 on the high schools. (ii) Shortly before, the Süddeutsche Zeitung had the headline "Jamaika-Sondierern läuft die Zeit davon" (Jamaica sounders runs out of time) (Süddeutsche Zeitung, 16.11.2017, p. 1). (iii) Another example is the award of the 2017 Nobel Prize for Medicine to US researchers Jeffry Hall, Michael Rosbach and Michel Young for their work on the day-night rhythm, the so-called 'inner clock'. (iv) As private customers in the mail order business, deliveries are offered in certain cities with a same-day delivery service. In Munich and Berlin, deliveries can be made after just a few hours.

Time also plays an important role for **companies**: companies have always been embedded in a value creation network: raw material manufacturers, raw material producers, parts and component manufacturers, end producers and dealers are the components of the supply chain. In earlier days, i.e. before the globalisation of supply chains, the majority of them were organised locally/regionally/ nationally and imports were only necessary due to raw materials or special parts. Today's supply chains are built globally thanks to global logistics and trade regulations. The competitive pressure of these open markets then led to the reduction of all capacity, inventory and time reserves, so that these supply chains are closely linked by serial interdependencies. In this respect, the term "supply chain" has been aptly chosen.

In the companies, the assembly lines are supplied "just-in-time" by the suppliers. This eliminates the need for warehousing at the purchasing company and reduces its costs. Only the time guarantee of the supplier guarantees the orders and prevents (in case of order) contractual penalties due to delivery delays.

The disaster of Fukushima or the discussions on hard Brexit are examples of the (possible) effects of "unplanned" disruptions of such efficient supply chains. In view of the withdrawal of the UK from the European Union, Japanese car manufacturers have announced that they will discontinue or review their production in the UK in order to ensure that production processes can continue.

With the digitalisation (i.e. industry 4.0, logistics 4.0 or purchasing 4.0), all participants now hope for further cost savings and acceleration of internal and inter-company processes. Previously unavailable data is now available faster or for the first time, enabling faster or more qualified decisions. The time benefit also becomes strategically important as an economic factor and enables new "time-based" business models.

However, digitisation is not uncontroversial. On February 28, 2019, Zeit-Online headlined the presentation of a new smartphone at the Mobile World Congress that supports the new high-speed standard 5G: "The Discovery of Speed" and chose the linguistic counterpart to Sten Nadolny's novel "The Discovery of Slowness"; a novel in which slow rhythm gives meaning to life. If this contradiction is now transferred back to 5G, the positive, i.e. meaningful, message of the new fast technology would be immediately questionable, i.e. faster technology would then no longer make sense. Was this the intention? This contradiction will be discussed later.

These wide-ranging examples are intended to show how ubiquitous the phenomenon of "time" is for all parties involved. In retrospect, this list could also be extended to include historical examples, e.g. the temporal rituals of indigenous peoples in harmony with nature or the development of the measurement of time using simple technical instruments. It is not possible to avoid "time"; time is omnipresent.

This book focuses on the business paradigms of time management and therefore it should be clear beforehand which aspects are **not considered**:

- The physical discussion at present (e.g. Rovelli, 2018): It is initially a fixed quantity as a result of the division of distance and speed. Albert Einstein formulates a "relative time" with general and special relativity theory and contradicted the statement that time is an absolute quantity. Thus, for example, the time elapses so differently in the course of a GPS measurement on Earth and in the GPS satellite due to the two speeds that a correction of time measurement must be made according to the Lorentz factor in order to compensate for positioning errors.

- The dimensions of space and time: Elias explains the four dimensions of space and time and pleads for a fifth dimension. This dimension should represent the personal experience, the consciousness or the personal experience. With this he wants to expand a theory of social symbols: Among its previous representatives of language/of meaning, space and time should also be included (Elias, 2017, p. XLVI , 26, 52 and 112).

- Heidegger's philosophical discussion about time: For him, the meaning of being does not only arise from the references to the present, but also from a temporal sequence, i.e. from the past (Heidegger's Gewesenheit) and the future.

- The field of tension between boredom, leisure, stress and amusement: This is where "flow" or idleness and positive or negative stress are discussed. (Luckner, 2012).

- The medical and psychological discussion about time, especially about the "inner clock": This phenomenon was investigated by Jeffrey Hall, Michael Rosbach and Michel Young and honoured with the Nobel Prize for Medicine. Their special focus was on the investigation of the day-night rhythm, i.e. the question of the mechanisms that control the 24-hour rhythm.

- The perception of time and the perception of time of people as personal and physical phenomena: The former refers to the length of time intervals that can be grasped by people with their senses. Only above a certain time threshold (duration) can facts be perceived by our sensory organs, e.g. perception of an optical stimulus by the eye or feeling of a heat stimulus on our skin. The perception of time, on the other hand, refers to the feeling of time, i.e. our evaluation of the actual duration of processes or processes in time. On the one hand, the intensity of mental activity plays a decisive role here. This feeling is expressed colloquially in phrases such as "time is running out, time flies by, time stands still, killing time". On the other hand, the perception of the chronological sequence can also be perceived in the sense of the "good old time" or "a turning point in time".

These aspects are not further elaborated in this book. This book deals with a special question for companies: the paradigms in time management. The term "paradigm" was chosen in order to express the fundamental, generally prevailing and no longer to be discussed opinion on predefined questions, here the temporal arrangement in enterprises. They reflect a generally accepted consensus on what solutions should be found for certain issues. In terms of time, the paradigm is that "acceleration" is the only direction underlying time management. It seems to be a kind of one-way street. In other words, ever faster means ever more successful or ever better. This is also evident from the above-mentioned introduction of the faster transmission standard 5G. Backhaus and Gruner (1997) speak early of an "epidemic of time competition".

Three special categories of time management and their respective paradigms are discussed in this book:

The first category considers time as a measure of efficient operational business processes: The operational processes form the backbone of the division of labour (services) and are to be managed (mastered) fast and reliably in the sense of efficiency. This means that time has an operative value.

The second category regards time as an expression of sustainable competitive strategy and as an opportunity to take a unique position towards competitors. Individual companies compete for more demanding customers with the aid of "time". In recent decades, globalization has facilitated or opened market access and supply chains have been reshaped as a result. Global competition has also become competition for time. In other words, time has a strategic value.

The third category considers time as a measure of the demarcation and characterization of societies. This time management of all people involved in a society does not pass our social developments by without a trace; on the contrary, it strongly shapes them. "Time" in the meaning of a modern spirit makes a clear mark on all of us. I.e., time has a social value.

The societal discussion seems justified because here too the shortening epitomizes the prevailing guideline. The result is a permanent feeling of haste and an existing time and deadline pressure. The link between companies and society takes places mostly through the employees, the buying behavior of the customers, the novelty levels of the products and the work organization.

The discussion of the three paradigms is prepared and thus easier to understand by first explaining essential phenomena and the functions of time. It will be shown that from the historical point of view the design of functions is a sign of our time. Therefore, in the basic chapter 1 the individual historical and cultural phenomena will be explained on the one hand and the functions of time will be explained on the other hand. Here it is a matter of explaining what is meant by the term "time" in view of its immateriality and how this view has developed (changed) in recent centuries. From these basics it will also be shown how the mentioned functions of "time" lead into the three categories of process, strategy and society.

In chapters 2 and 3, two functions are explained that are fundamental to the paradigms of time management: the standardization of language about time and time as an assessing criterion.

Chapters 4 and 5 focus on the third function, which is discussed on the basis of the entrepreneurial paradigm of time management. Acceleration is the predominant one-way street in time management. Chapter 4 deals with operational issues and chapter 5 with strategic issues.

Chapter 6 illustrates the permanent acceleration of social processes and thus their handling of the resource "time". Possible ways out (alternatives) of the never-ending acceleration of our social time are shown on the basis of time concepts. In this respect, this discussion could also be carried out from a pragmatic point of view. In this book, however, this takes place in two separate places: on the one hand in chapter 1 under the aspect of the analysis of social time structures and on the other hand in chapter 6 under the aspect of the possibility of escaping from this time vortex of society.

Each chapter begins with a well-known citation, which is intended to express its basic message briefly and concisely. In order to maintain legibility, the use of double forms or other markings for female and male persons is renounced.

I would be delighted if you as a reader would gain personal insights into the phenomenon of "time" and would not allow yourself to be drawn into the time vortex without reflection. For students as readers, I hope that they will gain a better understanding of the topic and sharpen their ability to judge temporal phenomena. For practitioners as readers, I hope to be able to create suggestions for their own operational discussion and its implementation.

I wish you lots of curiosity and sufficient "time" while reading.

Willi Darr

1. Time as an everyday phenomenon and its functions

a. Quote: Tomorrow is another day

This quote derives from the final sentence of the novel 'Gone with the Wind' by Margaret Mitchell.

This expresses that a new day begins after today and that the day is a social category of life. It is thus an everyday and repetitive phenomenon, since it is a universal form of systematization of the stages of life for human beings. This book does not discuss the extent to which the author wanted to address a new opportunity or a new beginning in the lives of the actors. However, on closer inspection, it is premature to conclude from the division into days, as written by Margaret Mitchell, that there is a perpetual and globally recognized categorization of the chronological sequence in the sense of a calendar. A brief overview of the basic understandings of time follows: a definition for this book, a historical and cultural retrospective of the phenomena, and the conception of the functions of time. This chapter is intended to make the rest of the explanations comprehensible to all readers and to enable each reader to better understand their individual understanding of time.

b. The definition of time in this book

The introduction to the foundations of time raises the question what time actually is, because it is not perceptible with human senses (touch, feel, taste, hear, see). Time also cannot be stored or recognized as a physical product. However, daily processes and events of life are perceivable.

The definition of time on which this work is based is now derived from the relationship between processes/events and a social symbol or institution that can be used by all. This relation of events with a social symbol is called "time" (see also Elias, 2017, p. XVII-XVIII). It thus corresponds to a created model (artefact) of the human being in order to fulfil certain social functions of a human being, a group or the whole society. The individual functions will be discussed in detail in section 1f.). Figure 1.1 shows a first example of this "setting in relation".

Here, the comparison of the duration or the individual points in time is mentioned in advance as an example of an evaluative function: The relationship-setting is outlined using the example of four activities (A to D) that take place one after the other. Here, the results of the preliminary stage should be the necessary input of the subsequent stage. The first activity A begins at time t_1 and the last activity D ends at time t_5. This establishes a relative relationship between the individual points in time of each activity (when does something happen?) and its duration (how many periods of time does something take?):

-   The beginning of activity A (t_1) takes place before t_2, the beginning of B (t_2) takes place before t_3, the beginning of C (t_3) takes place before t_4 and the beginning of D (t_4) takes place before t_5: $t_1 < t_2 < t_3 < t_4 < t_5$. Time t_1 is dated February 09, 2019 at 8:00 a.m.

-   The duration of A is the period from t_1 to t_2 (dA); corresponding statements for B, C and D are: d_B, d_C and d_D. The duration of d_A is shorter than d_B, which is shorter than d_C and which is shorter than d_D: $d_A < d_B < d_C < d_D$. Activity A lasts 2 hours.

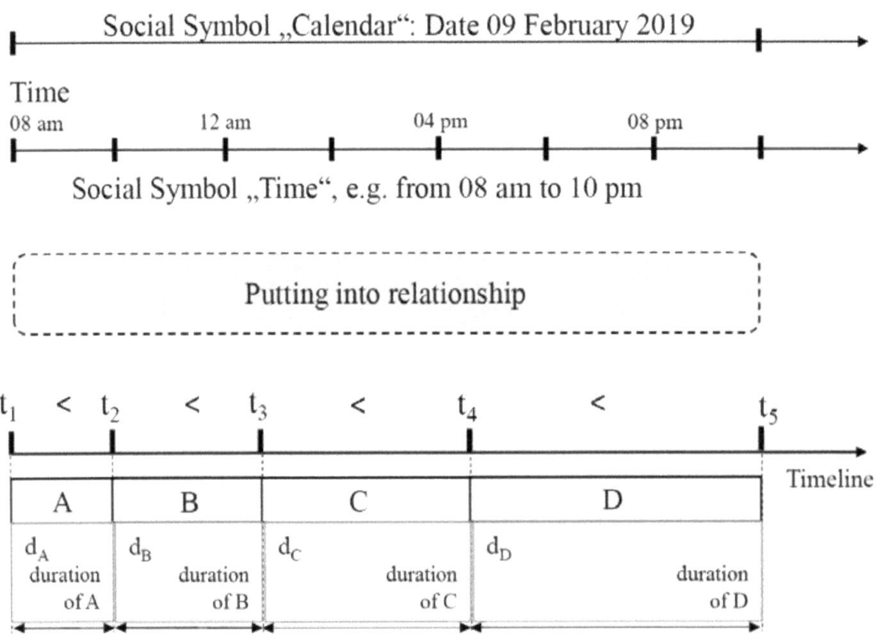

Figure 1.1: Relating activities

This social symbol corresponds (or better: corresponded at the time of their definition) to the rhythm of nature, recognizable for all, which humans could determine by simple observation at the orbit of the sun (seasons) and the moon (course of day and night). It was always possible to determine and ensure activities in harmony with nature through the two natural phenomena. This was done in the past and is now done using the clock and calendar. Thus Elias (2017, p. XXII) writes: "The man-made symbols of the changing dials of clocks [...] *are* time" [Fn1]. Strictly speaking, the calendar is still missing from Elias' statement.

The references of life to nature (sun and moon) are today no longer noticeable or necessary to this extent: The worldwide organisation of work and digitalisation permit a working world in 24-hour cycles and are no longer tied to the time of day. Thus, the 3-shift operation in industrial production and the 24-hour shop opening are established in trade. Additionally, production in the factories will be separated from months and seasons. This topic will be dealt with in more detail in Chapter 2.

The dials in the quotation from Elias (and the calendars to be added) are the concrete social symbol mentioned above. Time is the invisible relationship between the observable and the social symbol accepted by all, represented by the hours of the clock and the days/months/years in the calendars. Other forms will be mentioned later. These two symbols, the clock and the calendar, are only valid for all members of society (person, family, village, city, country, continent, world) through general acceptance. This process will be discussed in section 1c.).

The social symbol of "time" has in its origin derived from the natural rhythm of nature (the duration of the rotation of the earth in relation to sun and moon). Other possible artefacts of the evaluation of activities for the fulfilment of functions in a society were hardly possible in view of the means possible at that time, e.g. area-wide information for all participants, clear statement of the symbol or small technical employment.

The time in this definition can refer to all possible and conceivable processes or events of the world. Therefore, time can be classified as something that exists independently of social norms or specific events. Time has universal character and thus has the same character as the universal numbers in mathematics. Both concepts are ontologically neutral and get their meaning only by interpretation; regarding time this is done by "relating" everyday facts to a point in time or a duration.

The dials show the 24 hours for one day (or 12 hours twice) with 60 minutes each and 60 seconds each. The 24 hours are based on the astronomical hour count and divide the day into 24 parts independent of the season (equinoctial hour). This is based on the historical measurement and division of day and night of the Babylonians by the stars in the sky, i.e. 12 stars for the day and 12 stars for the night. They also made their calculations in the sexagesimal calculation system that is commonly used today; as well as the 360-degree division of a circle. The divisions of leap years, leap days and leap seconds will be discussed later. The days of the individual months are based on the harmony of the time divisions with the natural processes of the sun and moon. The daily hours are also determined differently in Europe in summer and winter time.

Detached from the lunar or solar calendars, other temporal divisions can or could exist for humans, which are only dealt with in the context of a list (see in particular Horx, 2003, p. 65ff.). These do not form the basis of a universally accepted scale of time in our societies and do not have the functional power of a clock or calendar. These are the following categories of time:

- Aeons: These are changes and evolutionary developmental rhythms in nature, e.g. climatic leaps.

- Types of civilization: These are social models, e.g. the hunter-gatherer, the agrarian civilization, the industrial civilization and the knowledge economy.

- Contratieff waves: These are the relatively constant technological waves, e.g. steam engine/cotton, railway/steel, electricity/ chemistry, car/oil, computer/ information, genetic engineering and biotechnology, nanoengineering and sideral technology.

- Business cycles: These are the patterns of business cycles and recessions.

- Megatrends: These are long-term, interdisciplinary developments valid all over the world. John Naisbitt coined this term with his book Megatrends. Examples include individualization, globalization and digitalization.

- Zeitgeist: These are the ups and downs of the individual market segments and markets, e.g. telecommunications, discounters or shopping centres.

- Fashion trends: These are the developments of individual product categories, e.g. DVD, Italian espresso machines, miniskirts, wasp waist, smartphones or tablets.

The universal character of "time" on the basis of the natural recurring phenomena of the sun and the moon has established itself worldwide as the basis for establishing relationships. This is also explained by the cultural time of origin of this symbol.

Therefore, in the following it will be given a short review of the different forms and types of the divisions.

c. Selected phenomena of time: a historical and cultural overview

This social symbol, especially the calendar, and its basis have changed several times in retrospect. A short historical review should clarify this:

(i) In 46 BC Caesar initiated a draft of a year with 12 months, each with 30 days and five additional days at the end of the year. This proposal was transformed into a February with one day less and six additional days for the six odd months. Thus, this comes very close to today's seven months with 31 days and a February of only 28 days (in the leap year with 29 days). (Elias, 2017, p. 185). The first basis is thus the Julian solar calendar. The calendar year lasted 365.25 days.

(ii) Easter was fixed at the Council of Nicaea in 325 AD on the first Sunday after full moon, which follows the church-agreed beginning of spring (always March 21). (inter alia, Elias, 2017, p. 186). The astronomical beginning of spring, however, is in the northern hemisphere between the 19th and 22nd of March. In 2019 the Astronomical Spring Equinox was on March 20 and full moon on March 21 was still assigned to winter, so that April 19 was counted as the first full moon. Easter Sunday was therefore on April 21, 2019.

(iii) King Charles IX of France set a uniform date for the beginning of the year with the Edict of Roussillon of 1564: from March to 1 January. As a result, the situation of the months changed: September from the 7th to the 9th month, October from the 8th to the 10th month, November from the 9th to the 11th month and December from the 10th to the 12th month, even if the names of the months then no longer fitted linguistically.

(iv) Pope Gregory XIII decided in 1582 to revise the Julian calendar. The Julian year is 11 minutes and 14 seconds too long compared to the solar year. Gregory XIII therefore omitted ten days in order to better adapt the existing calendar system to the physical conditions (i.e. the orbit of sun and moon) (Elias, 2017, p. 22f.). Therefore, this was replaced by the Gregorian solar calendar in the context of its reform, and the new year lasts 365.2425 days. This new calendar treats the temporal deviation from the solar year more flexibly, by also a correction cascade for the compensation of the deviation of the 365 days of a year from the sun process exists: leap second, leap year and leap year hundred. The changeover from the Julian to the Gregorian calendar was, however, delayed in the countries (see Time, 2018):

-   with 10 skipped days 1582 in France, Italy, Poland, Portugal and Spain, 1583 in Austria and Catholic Germany, 1587 in Hungary, 1610 in Germany (Prussia),

-   with 11 skipped days 1752 in Canada, Great Britain with colonies and the USA,

-   with 13 skipped days in 1916 in Bulgaria, 1918 in Estonia and Russia, 1923 in Greece and 1926/1927 in Turkey.

(v) The division of a year into twelve months and the allocation of a globally accepted year is not uniform. The basis can be the solar year (solar calendar), the moon cycle (lunar calendar) or a combination (lunisolar calendar). The year-counts differ in the different cultural circles:

-   The Christian world begins the calendar with Christ's birth and counts the years "before Christ's birth, BC" and "after Christ's birth, AD".

-   Russia, on the other hand, retained the Julian calendar and therefore shows a difference of 13 days to the Western European census: This is exemplified by the Christian Christmas (on 25 December) and the Orthodox Christmas (on 7 January). Even Easter takes place with a time delay: in 2019 Easter was celebrated in Russia on April 28th.

-   The Jewish calendar is a lunisolar calendar, i.e. a combination of a lunar calendar and a solar calendar, since the 12 lunar months (354 days) deviate from the solar year (365 days) and require a compensation regulation based on the solar year. The counting of this calendar also begins in the year 3761 BC. The Jewish day is calculated here from one evening beginning to the next evening.

-   In the Islamic world, the hijra calendar applies. It is a lunar calendar with twelve lunar months, which are about 11 days shorter than the solar year. As beginning of the Islamic time calculation serves the 1st Muharram of the year: according to Christian time calculation this was on 16 July 622 AD.

(vi) In 1793 the French National Assembly adopted a "revolutionary calendar": 1792 was therefore year 1, which began on 22 September; each month had 30 days with five extra days at the end of the year; the week had 10 days; each day had only 10 sections (not 24 hours); each section was measured in decimal (no longer in 60 sections). This calendar was abolished after 13 years. (Levine, 2016, p. 118f.).

(vii) In 1929, Stalin also introduced a "revolutionary calendar" in Russia: the week with five (later six) days and a month with six weeks. This calendar ended in 1940 (Levine, 2016, p. 119).

(viii) Also, in recent time changes of timing are made: Since April 5, 2018 (here: 2018 from the European calendar) North and South Korea (again) have the same time.

As a result, the different countries and the different cultures also had and have different bases, counts, classifications, equalization regulations and, as a result, different national or religious festive days and holidays. The determination of the social rhythm was therefore also a social agreement and control (Levine, 2016, p. 118f.).

d. Selected cultural studies on time

Time and its phenomena are not only structured differently in different cultural circles but are also perceived differently in different cultures. This has been investigated several times. Three well-known studies are presented here:

(1) **Edward T. Hall** published in 1983 in his book "The Dance of Life: The Other Dimension of Time" different types of time and time perception. Two statements underline his basic conviction:

(i) „*My goal in this book is to use time as a means of gaining insight into culture, but not the reverse.*" (Hall, 1983, p. 5) and (ii) „*Without people, technology means nothing. If the world's problem are to be solved, it will be by human beings, not by machines; the machines are only here to help us.*" (Hall, 1983, p. 9).

In order to achieve his goals, Hall then developed eight different types of time in form of a "Map of Time", which are spanned by the poles "cultural or physical time" and "individual or group time". Thus, Hall develops a square with the corner points (i) the "philosophical & conscious time", (ii) the "unconscious pressing time", (iii) the "content-rich time" (situational culture) and (iv) the "explicit technical time" (content-empty time). On this basis he then distinguishes the types of time (i) metaphysical time, (ii) sacred time, (iii) profane time, (iv) micro time, (v) synchronous time, (vi) personal time, (vii) biological time, and (viii) physical time.

Hall then focuses on a division of time perception: the polychronic time (P-time) and the monochrome time (M-time). A first delimitation of Hall leads to the following statement:

„Years of exposure to other cultures demonstrated that complex societies organize time in at least two different ways: events scheduled as separate items - one thing at a time - as in North Europe, or following the Mediterranean model of involvement in several things at once." (Hall, 1983, p. 45).

In doing so, he focuses in particular on the appreciation of the importance of time in terms of punctuality in different cultures: A monochronic sense of time represents thinking in linear sequences of activities (horizontal). Everything happens one after the other. As a result, punctuality has a high value in these cultural circles, as otherwise all subsequent processes would be delayed. In order to avoid waiting times, all activities (including personal life) must be coordinated and planned. The priorities of the alternatives determine the sequence of activities and thus save time, i.e. time is not lost, time is not killed, or time does not run out. Tasks, processes and plans are at the centre of M-time.

In contrast, polychronic time perceptions are based on a cyclic and simultaneous time sequence (vertical). Everything happens simultaneously and the punctuality of an activity is not considered relevant. Waiting times of individuals (and larger waiting rooms in public buildings) are the logical consequence (from the point of view of the M-world). The role of friends who are given time is far more important in order to always be up to date in all questions taking place in parallel. The individual people (friends) are at the centre of P-time. The discussion of individual countries and their habits and perceptions of time is not dealt with in detail here.

(2) Secondly, the cultural studies of **Trompenaars/Hampden-Turner** are presented in their book "Riding the Waves of Culture. Understanding Cultural Diversity in Business" from 1998: Her studies currently examine the different attitudes on the basis of seven cultural dimensions. In the section "How we manage time" (Trompenaars/ Hampden-Turner, 1998, p. 120ff.), they examine the different ways of perceiving time. As a result, three categories are distinguished which can classify the cultural perception of time:

- Time orientation can be past, present or future oriented, i.e. countries with a strong preservation of traditions, with a life in the present moment and a strong focus on future projects/ideas can be characteristic. The two researchers presented the characteristics of time orientation for the individual countries examined in the form of three circles (Trompenaars/Hampden-Turner, 1998, p. 127), which express the above three characteristics in their relative weight and the connections between past, present and future. This is a possible measurement of the cultural profile for the time orientation of individual countries. This method ("Circle Test") is the work of Tom Cottle (cited by Trompenaars/ Hampden-Turner, 1998, p. 125).

For example, Russia or Venezuela are assigned to the past culture, Germany to the present culture and Korea to the future culture. In China or Mexico, the sense of time is regarded as balanced. The three sensations are strongly overlapping in Japan or Malaysia, connected in Northern Europe and unconnected in China, Mexico and Venezuela.

- The structuring of time can be based on a sequential or a synchronous concept of time, i.e. time is either planned as a sequence, appointments are made/kept (sequentially) or time is a cycle of several things and success depends on people's relationships and not on their punctuality (synchronously). These studies coincide with those of Hall regarding the monochronic/polychronic culture, i.e. M-time or P-time.

- Thirdly, they examined the long-term or short-term nature of their own plans or thoughts. Thus, the time horizon of the past or the future considered was measured for the individual countries. Examples include the short-term orientation of the Americans or the long-term orientation of the Chinese for the past and the future (Trompenaars/Hampden-Turner, 1998, p. 129). For interested readers, Trompenaars and Hampden-Turner have listed information on conversation and behaviour for the individual time types on pages 138 to 140 of their book.

(3) In his book from the years 1999 (first edition) and 2016 (20th edition) "A map of the time. How cultures handle time " **Robert Levine** demonstrates that the social sense of time in a cultural circle has deep consequences for the psychological, physical and emotional well-being of an individual.

He first distinguishes between the three types of time "clock time", "nature time" (rhythm of sun and seasons) and "event time". In the event time (Levine, 1997 and 2016, p. 122ff.) there are no clocks and no calendars. Instead, activities are triggered by events (including those of nature): e.g. fatigue, hunger, tides or the position of the sun determine subsequent activities. This philosophy of life is therefore expressed in idioms such as "giving time to time" or "time needs time" (Levine, 1997 and 2016, p. 127). Known examples are also the dry season, the rainy season or the mealtimes of a new born child.

In clock time, the clock or calendar defines the start and end of an activity. The developments in the measurement of time and the standardisation of time or the development of time targets will be discussed later. Nature-time is initially based on the natural rhythm of the sun and moon. Levine (1997 and 2016, p. 99) also mentions water levels and the occurrence of the ringlet worm as the beginning of the planting season.

He examined 31 different countries with regard to their respective speed of life. In Part I of his book (Social time. The heartbeat of culture), he presents a measurement method for social time using the following three indicators: (i) the average speed of randomly selected pedestrians over a distance of 20 metres, (ii) the time it took an employee at a post office counter to sell a standard stamp, and (iii) the accuracy of 15 randomly selected clocks at bank buildings. On this basis, he could measure the pace of a society. He attributed the different temporal characteristics to five basic factors of cultures on earth: (i) the prosperity of the economy (ii) the degree of industrialization, (iii) the number of inhabitants, (iv) the warmth of the climate, and (v) the individualism of cultural orientation (Levine, 2016, p. 37f.).

In Part II of his book (Fast, slow, and the quality of life) he presents his results for the 31 countries with regard to the pace of life. An overview is printed in his book of 2016 on page 180: the eight fastest countries come from Western Europe and the slowest countries are the non-industrialized countries in Africa, Asia, the Middle East and Latin America (Levine, 2016, p. 180ff.). There are also significant differences within a country, e.g. in the United States (see the overview of individual cities in Levine, 2016, p. 200f.). Interested readers are referred to Levine's comments (1997, p. 153ff.) on the connections between cultural time and health, wealth, happiness and social commitment.

These remarks on the historical review and on the various cultural studies show that **the** time, i.e. the everlasting social symbol valid for all, did not exist despite the clear definition at the beginning of the chapter, but was subject to several changes in content. It seemed helpful and necessary to explain this review and these cultural anchorages[Fn2]. It becomes clear that the definition of time in different societies shows

16

enormous differences. In the following, the social development of time in Western societies will be clarified.

e. Analysis of social time structures: The concept of Acceleration by Hartmut Rosa

In his book "Acceleration" (German: Beschleunigung), Hartmut Rosa put up for discussion a much-noticed social analysis based on time (Rosa, 2016a). In doing so, he does not compare the temporal differences between different cultures of Hall, Trompenaars/ Hampden-Turner or Levine, but instead he undertakes a comparative analysis of generations of Western culture, especially premodern/early modern, classical modern and late modern. Therefore, he first designs an explanatory model and then applies it as a prognosis model.

His approach assumes that (i) time is fundamentally a key category in the analysis of the social world (Rosa, p. 19) and (ii) social developments in particular can be distinguished from one another and identified on the basis of structural changes in time structures and time horizons. He expresses the expression of his temporal measurement results as "social acceleration" (Rosa, p. 24). In the following, the citation method in this section is limited to the author and the page number, since only this one source (2016a) of Hartmut Rosa is cited.

The time of each individual in a society is determined by the time of everyday life (e.g. daily work, leisure, sleeping), of life (e.g. studies, working hours, retirement, parenthood) and of our epoch. These three-time levels together determine the "being in time" of each individual (Rosa, p. 30f.).

The focus of his work are the dimensions of acceleration (Rosa, p. 124ff.).
- The technical acceleration (his chapter 4)
- The acceleration of social change (his chapter 5)
- The acceleration of pace of life (his chapter 6)

Rosa does not ignore the fact that societies also have retarding forces (decelerations). Thus, it represents the following categories of such insistances: the natural speed limits (e.g. physical limits or the seasons), deceleration islands (e.g. sects, wellness oases), deceleration as a side effect (e.g. traffic jams, unemployment) and the intended deceleration (e.g. ideology or preparations for the next acceleration thrust), (Rosa, p. 138ff.).

The technical acceleration refers directly to the global value chains and focuses on the faster transport, picking and manufacturing processes. Globalisation and developments in information and communication technologies have led to the following effects:

- People, data and goods are moved with increasing speed, and goods are produced in accelerated manner.

- Social events become "placeless" because, for example, the location is no longer visible on the Internet.

- The world is brought to man (i.e. the goods of the world are made accessible from everywhere) and no longer man is brought to the goods. Virillo invented the term "furious standstill", i.e. the standstill of people and racing processes of goods provision in their environment.

As a result, he notes from the technical acceleration a change in spatial, social and thing relations (Rosa, p. 170ff.). In particular the accelerated social change and the pace of life on the basis of technical new basic conditions are shown by Rosa. Thus, he comes to the same conclusions as Postman (1992): New technical developments always have social consequences.

In accelerating social change, the focus is on the shift from intergenerational to generational to intragenerational tempo (Rosa, p. 179). He makes this clear in the following key points:

- *Family*: from a family "until death do us part" over life stage partners with divorce rates to the "serial monogamy" and/or "the lovers on time" (Rosa, p. 180f.)

- *Profession*: from a family dynasty with over-generational stability over its own independent choice of occupation that creates identity to frequent changes of occupation and jobs in part-time or temporary work. The latter is accompanied by a high risk of dismissal or an expected short period of employment.

- *Generations*: from the almost immovable position in society over the formable role in society to the liquefaction of social and material relationships (e.g. friends, property, insurance, places of residence).

As a result, Rosa concludes that post-modern society can be characterized by the "slippery-slope phenomenon", in which all areas of human life are on a slippery slope, as it were, and standing still becomes impossible through preserving/not acting (Rosa, p. 190). The question of the next steps in a person's life (the question did not arise in pre/early modernity or "classical" modernity as in post-modernity) can then only be assessable and decidable from the moment. The best-before date of each condition is close to zero. In effect, stress and time pressure can then be determined in two respects, since the individual must keep pace with his environment and, in view of the total temporalization, must also be on the permanent search for the next follow-up option. (Rosa, p. 189ff.).

He does not examine the acceleration of the pace of life, the third area in Rosa's conception, on the basis of the three indicators of Levine (1997), but rather defines the increase of the episodes of action and/or experience per unit of time as a criterion (Rosa, p. 198), which he then divides into objective and subjective components.

In the objective sense, acceleration occurs when the activity is carried out more quickly, when idle times between activities are avoided, when activities are carried out in parallel or when other, faster, activities are carried out (Rosa, p. 199). This is demonstrated by studies or observable phenomena such as power-nap, speed-dating, the duration of commercials, the average time for eating, the zapping of TV channels or the response time of messages received, particularly in social media. This objective acceleration increases c.p.'s planning and time requirements for personal coordination and synchronization of personal daily actions.

The objective increase in speed then leads to a faster subjective experience of time (Rosa, p. 213ff.). His initial hypothesis is that with the increase in pace of life, time resources become scarcer, i.e. the feeling says one has less time or time pressure has increased (Rosa, p. 214). He attributes this to the increased fear of missing out and the increased pressure to adapt. Every reader should check for himself to what extent the smartphone as a permanent companion is an indication of this. In its extreme form, the sense of time is characterized by a short-short pattern (short experience time and short memory time) or a long-short pattern (long experience time and short memory time). Thus, even with an objectively longer leisure time, there is nothing left of experience, i.e. one's own experienced history, is not integrated into one's own tradition and thus not meaningfully linked into a relationship. The past experience is recalled (without experience) and decontextualized, although society is rich in experiences (Rosa, pp. 223ff., especially pp. 230ff.).

The following applies to personal self-understanding: "Who you are is always determined by how you became it, what you were and could have been and what you will be and want to be". (Rosa, p. 237). Against the background of technical acceleration, social acceleration and the acceleration of pace of life, Rosa draws the conclusion that personal identity in the extreme shrinks to a "punctiform self" (Rosa,

p. 238). Using the example of the baker, this means that one is not a baker, but is currently working as a baker.

He attributes this development to the following reasons (Rosa, p. 241ff.):

- The three forms of acceleration reinforce each other.
- Time is an economic value that is only preserved if it is permanently shortened.
- Only an accelerated life increases one's own self-confirmation in comparison to one's neighbour or competitor.

In Rosa's final consequence, this leads to hyper accelerated societies in which things no longer have any meaning, the past is not a reference and the future is perceived as meaningless. He draws a conclusion in the form of a crisis situation in which there is nothing permanent to decide (Rosa, p. 420ff.). He speaks of the "simultaneous time" (Rosa, p. 347) or the "timeless time" (Rosa, p. 344). Life plans that strive for stability or other fixed points of life are stamped as doomed to failure.

	Premodern/ early modern	Classical modern	Late modern
Key messages	Work and family are stable over generations Static historical perspective Given identity of each individual in society (a priori)	Starting a family and choosing your own career The temporalization of history Self-determined identity in comparison to the family ancestors	Jobs and life stage partners change more frequently Simultaneous time Hyper accelerated society without a reference-creating past

Table 1.1: Selected statements on the classification of the societies

Some of his key messages are illustrated in Table 1.1. His concept, which he proves by comparing with a large number of studies, shows clearly how a stable society has changed in premodern/early modern times as a result of a cycle of self-reinforcing (technical) developments. This passive role of "being changed" expresses this "slippery-slope phenomenon". Rosa fixes the personal perception of time to the three dimensions of social acceleration. He thus carves out that technical developments have a formative effect on the social and cultural sphere of human beings. Postman built up a parallel argument by pointing out the transition from Tool Culture via Technocracy to the Technolpol: "Under the rule of the Technolpol we are urged to spend our lives striving for "access" to information. (Postman, 1992, p. 70). He notes the collapse of defensive mechanisms in dealing with the wealth of information, the lack of tradition[Fn3], ethical principles and, ultimately, the emptying

20

of symbols. Rosa consistently refers these consequences to the individual temporal dimensions and Postman to his conceptual components.

As a brief conclusion, it should be noted that time in recent centuries has been a significant attribute in the characterisation of (western) societies and that technical and social conditions have always been related to "time".

f.   Time responsibilities and measurement methods

The competence of time, and this is no surprise in view of its social significance, was and is up to the national authorities: in Greek mythology Chronos was a god, i.e. the god of time. Later, time was determined by priests and kings, and time was always the monopoly of the state (see Elias, 2017, p. 19f.).

A uniform time determination for the German Empire was previously put into effect with the Central European Time on 1 April 1893 (law concerning the introduction of a uniform time determination). It corresponds to the mean solar time on the longitude 15° East and has a deviation of one hour from the so-called coordinated world time (UTC+1) or in summer time of plus two hours (UTC+2). The introduction of these European time zones in the individual countries took place in the years after 1893, but with a time shift by country.

The current legal basis for the jurisdiction of time determination in the Federal Republic of Germany is regulated in Article 73, paragraph 1, no. 4 of the Basic Law: "The Federation has exclusive jurisdiction over the legislation: [...] the determination of time."

Until July 11, 2008, the Federal Republic of Germany applied the Act on Time Determination (Zeitgesetz, ZeitG). It defined the legal time in official and business transactions as decisive with regard to the coordinated world time (Central European time plus one hour), the second as the base unit, the leap second, the authorisation to introduce Central European Summer Time and the responsibility of the Physikalisch-Technische Bundesanstalt (PTB: Physically Technical Federal Institution) to measure and disseminate the legal time. With effect from 12 July 2008, their contents were incorporated into the Act on Units in Metrology and Time Determination (Einheit- und Zeitgesetz, EinhZeitG).

Since 1969, time in Germany has been officially measured with the help of the atomic clock in Braunschweig at the Physikalisch-Technische Bundesanstalt. Caesium atoms act as clock generators.

A brief glance at the various methods used to measure time, for which several methods have been used over the centuries, is enough to illustrate this: A very original measurement of the time of day is made using the position of the sun with the help of sundials to determine the time of day. For the determination of the months, the phases of the moon, the solstice, the lunar and solar eclipses respectively the new moon were used (Elias, 2017, p. 63ff. and Levine, 1997). Stonehenge is also considered such an example of a cult site with a built-in time scale (Elias, 2017, p. 62f.). The Romans used water clocks. The cooking time of the potato was used to measure time in Peru (see Potato, 2018). Sand timers were also used early for measurement. Around 1700 pendulum clocks were developed and from 1850 the first wristwatches (see the illustrations in Levine, 1997 and Elias, 2017).

The statements have intentionally been kept brief and are intended to give a glimpse into the technical diversity of the measurement of "time".

g. Functions of time

The explanations given so far now raise the question of why people actually needed and need time determinations. Would social life be possible without time measurements and without time management? And what are their benefits? So, a connecting link between the individual members of society or today the companies in the global supply chain, their departments or their employees is needed. Time thus becomes an instrument for the coordination of all the interlinking facts of a society and its instruments are expressed in several functions. The own personal behaviour and feelings can then be understood, classified and, if necessary, re-evaluated with help of the time signals sent. This also makes it possible to categorize and prioritize the common challenges and problems that people seek to overcome with help of "time" (Elias, 2017, S.XVIII, XXI and 76).

In this book the following categorization of the functions of time is used: There are the three levels of **semiotics**. They are clearly delimited and successive stages and allow a non-overlapping representation of the individual functions. A distinction is made between the three categories (i) syntax, (ii) semantics and (iii) pragmatics, which date back to Charles Morris (1938). He dealt conceptually with the theory of language (see also Picot et al., 2003, p. 89ff.).

The syntax (Morris, 1938, p. 13ff.) explains the characters and symbols used in a language. It is the prerequisite for mutual understanding, especially for operational processes to be organized between companies. This means that the identical syntax forms the common basis of the language in time management. This is explained in Chapter 2.

The semantics (Morris, 1938, p. 21ff.) is dedicated to the meaning of the individual signs and statements and evaluates, for example, the operational processes on the basis of the common language. It gives those who are responsible an orientation by comparing the time situation between the partners in the supply chain and assessing their own position against that of their counterparts. In other words, semantics provides the criteria for the evaluations for time management. This is explained in Chapter 3.

Pragmatics (Morris, 1938, p. 29ff.) focuses on the meanings of temporal statements and evaluations. It discusses the meaning and purpose of the whole message. This book focuses on the decisions in the supply chain which, according to the basic statement, are made paradigmatically and show a strict uniformity in their basic direction. The operational and strategic decisions are assessed and decided exclusively on the basis of time. This is explained in chapters 4 and 5. The societal discussion on time management takes place in chapter 6.

The discussion of the functions of time has also been done by Elias and discussed in several sections of his book, without making it his core of thought. In several chapters, he has also highlighted three essential functions (Elias, 2017, p. XIX, XLV and 19ff.) for forming relationships for human beings: he speaks of the function of communication, the function of orientation and the function of regulating human behavior. For him, time has a clear instrumental character, too.

The first function of time, communication, is thus the common language and a necessary means of effective cooperation (socialization) of all members of society (Elias, 2017, p. XXI, XXVIII and XXIX). Without a common agreement on the language (here: time), cooperation in the community is not possible or only possible with great effort. Time is a common language with which the exchange of messages is carried out and which is also understood by the other person. This function (communication) presupposes the signs, the meaning, their transmission and their understanding. Only then can the communication process take place as such. The second function, orientation, is a common evaluation yardstick for the individual activities, so that each individual can evaluate the own behaviour in a comparative way. With time as a yardstick, individual activities can be related. The third function, regulation, refers to the normative value concept of the community, i.e. the desired target states and individual deviations are defined with the third function mentioned. This comes very close to orientation. In terms of content, individual decisions are coordinated with regulation or instructed in a hierarchical process. Ultimately, effective decisions affecting several participants in society are the reasons for regulation.

In summary, the following three functions of time are considered and represent the core of this book:

(i)　Function of the common sign/language (syntax). For Elias, this corresponds to communication. This is the first basis of the pragmatic discussion.

(ii)　Function of evaluation (semantics). This corresponds to orientation in Elias. This is the second basis of the pragmatic discussion.

(iii)　Function of operational, strategic and social decisions (pragmatics). For Elias, this corresponds to regulation.

Thus, the three functions can also be used to justify the focal points of this book:

- As the leading authority of a company, management also sets the value standards. Of course, these include the roles of the processes. But the limits to intensively pursued time reductions also become clear in the context of risk management, as the robustness of the organization and thus the longer-term safeguarding are directly influenced. This discussion has an impact on operational and strategic issues.

- For companies, the management of time is always also a management of operational and inter-company operational processes. Hence the explanations on the current paradigms-in operative management.

- Thirdly, strategic management evaluates the possibilities companies can have by time-positioning on the customer market to add value for the customers (time as value and/or as product) and better than the competition. Hence the explanations to the paradigms at present in strategic management.

The longer and more differentiated the global value chains become, the closer this time-related discussion becomes relevant to top management decisions. The serial, pooled and reciprocal interdependencies become increasingly relevant and the "regiment of clocks" becomes even stricter (Elias, 2017, p. 99). An unstable and complex network of relationships in the supply chain or in society makes time management urgent and necessary. Companies in such a framework will receive an immense amount of time signals and will in fact be forced to handle them (perceive, evaluate, control, regulate). In simpler network constellations, e.g. local markets, the time signals can be manageable and a simpler form of management is sufficient. An example of a simple situation is the Kanban supply replenishment, which contains the rule "empty box, then deliver one box", and is suitable for stable or simple chains. It is emphasized here that the development of big data or artificial intelligence also seems to be a management necessity against the background of unstable and complex networks, because the simple rules are then no longer powerful enough to effectively control the challenges.

The need for further development of companies in these unstable and complex situations is expressed in the need for increased communication (additional and timely collection of data), increased orientation (additional and timely processing of data in sense of obtaining information) and increased coordination (preparation and triggering of decisions).

In summary, Figure 1.2 shows the various key messages that developed from life in harmony with nature to the coordinated industrialised world. They form the historical background of the representation and discussion of the following three functions of time.

Different historical developments of peoples or countries
 	
From living in harmony with nature to the coordinated industrialized world
	• Calendar in alignment with the orbit of sun and moon
	• Further developments of calendars
	• Culturally different calendars
	• Development of watches according to technical possibilities (hours, minutes, seconds)
	• Culturally different time assessments:
	  - P-time/ M-time
	  - Nature Time/Event Time/Clock Time
	  - Reference to the past, present and/or future
	  - Long-term/short-term thinking
	  - Socially accelerated developments

Functions of time		
Syntax (characters)	Semantics (assessment and target dimensions of time)	Pragmatics (sense and operational, strategic and social discussion)

Figure 1.2: Key statements on the phenomenon and functions of "time"

2. Syntax: Standardization of the language of time
a. Quote: 20 equals 15

Behind this mathematical statement is the question of common understanding and a common language. Mathematically, 20 is greater than 15, and the statement "20 equals 15" is therefore wrong. However, this conclusion is not self-evident. A judgment of the Federal Constitutional Court provides an example:

The background to this quotation goes back to a decision of the Federal Constitutional Court in 1985 in which the duration of the basic military service (at that time 15 months) and the duration of the alternative service (at that time 20 months) were in relation to the decision. The interested reader can read the judgment under BVerfG of 24.04.1985, file no. 2 BvF 2/83.

The discussion becomes inflamed in view of Article 12a paragraph 2 sentence 2 of the Basic Law: "The duration of alternative service may not exceed the duration of military service". Accordingly, 15 months (basic military service and alternative service) would have to apply to both services, in short 15 equals 15.

Nevertheless, the Federal Constitutional Court has decided that the 20 months of alternative service do not exceed the 15 months of military service, i.e. 20 = 15. For example, recital 65 states: "This does not contradict that civil service often lasts longer than military service, which in the past on average actually had to be completed. For in practice, the legally permissible maximum periods for military exercises have not yet been regularly taken up [...]. Continue in para. 66: "Again, the given differences between military and civilian service must be taken into account: The person doing community service performs his or her service contiguously and conclusively, is generally subject to a less strict employment relationship and is typically in a less stressful situation.

The conclusion of the argumentation is that 15 months of military service plus "potential exercises" plus "less severe/burdensome" corresponds to 20 months of alternative service.

Judges Mahrenholz and Böckenförde disagreed: In recital 150 they say: ""Duration" is a quantitative term that refers to periods of time". And further in recital 151: "The Basic Law does not speak of "burden", but of "duration". The concept of burden is not suitable for the interpretation of Article 12a (2) of the Basic Law if it overrides the concept of "duration of service", but only if it submits to it.

"Equal burden" in the sense of Article 12a (2) sentence 2 of the Basic Law can therefore always only mean: "Equal burden in the duration of service". The provision does not therefore permit "given differences between military and civilian service" in the sense of a greater burden of military service than civilian service to be taken into account when determining the duration of civilian service."

In view of the two conflicting opinions, 20 months is longer than 15 months; in view of the judgment, 20 months does not exceed 15 months.

This description illustrates that temporal terms (here: duration) do not obviously express the same facts for all parties involved. The different historical and cultural calendars that refer to different bases are recalled here, although the common base is the sun and the moon respectively. A common language about "time" is therefore not self-evident, since (in the words of the above judgment) "incriminating and strict" has been taken into account.

This common language without further implicit premises can be attached to the clock as a universal social symbol and therefore plays a central role. It shows a recurring sequence pattern and then serves as a standardized reference pattern for a second sequence of events (Elias, 2017, pp. XVII and 97). Clocks de facto show only an identical duration between two cycles. But clocks (and calendars) have an instrumental character and are means to an end. They fulfil this purpose, i.e. they become the social standard, because the second, minute, hour, day, month and year are standardized and accepted in their duration for all people who want to coordinate. To the question "What exactly do clocks indicate? (Elias, 2017, p. 95) then follows the answer "time". This does not only mean the duration for a single individual, but for society as a whole.

In addition, time determination is based on the human ability to link several processes, one of which serves as a yardstick for the other (see also Elias, 2017, p. XXIII and 42). This common language is all the more urgent, as multi-faceted and complex the temporal circumstances are: The relationship between the daily processes and the recurring process of a clock leads to a comparison of earlier and subsequent processes. The main key of time determination lies in the specific ability of people to grasp a continuous sequence of earlier/later or before/after and to link it via "time", i.e. the time or calendar. Consequently, terms such as "sooner" or "later" are also an expression of our ability to imagine together what does not happen together and what we experience as not happening together. This applies to the past, the present and the future (Elias, 2017, p. 44-46). If these statements of Elias are now applied to global and culturally different time understandings, e.g. M-time, P-time or present and/or past related, the misunderstanding is pre-programmed as with the Tower of Babel. For global cooperation it needs an identical syntax, i.e. a global language about "time".

A first possibility for forming the common syntax is to create this common language through cultural learning. However, this is opposed by the fact that the current and cultural sense of time is deeply rooted in human beings. Elias quotes Hall's studies, which clearly compare the time assessments of Pueblo Indians and American employees (Elias, 2017, p. 118ff.). It takes seven to nine years for an adolescent to learn the distinct sense of time learned by a cultural group (cf. Elias, 2017, p. 120). This common sense of time shapes human relationships of all kinds. These would be severely impaired and could hardly be sustained in the long term if one's own behavior were no longer coordinated or regulated according to a collective time scheme. In this respect, the determination of time (i.e. the formation of a relationship between event and social symbol) and the social regulation of everyone must first be learned (Elias, 2017, p. 121), because first of all there is a gap between "individual" and "society" with regard to language at present and society as a whole is hindered in the sense of further development (Elias, 2017, p. 123). This fundamental learning of the perception of time has developed differently over the centuries in different countries and is thus ruled out as a short-term possibility.

There is a second possibility of a common language: in his studies, Elias has particularly considered the evolution of human history from the theocentric universe and has found that "time" has been replaced by natural laws and increasingly mathematized: The "time" represents a physical movement, i.e. a "physical time". The emergence of "physical time" from the matrix of "social time" and the emergence of a dualism of the concept of time from "physical time" and "social time" has occupied him very much (Elias, 2017, pp. 85 and 91-93). A second time level is de facto opened.

Thus, this second possibility for the formation of a common syntax can be assumed in the following, i.e. a parallel language to "time", which basically does not contradict social time. This possibility will be explained in more detail below by highlighting the individual stages of this standardization.

b. A review of standardization of time

The standardization of time as a common language (here: the different local times) was only carried out later: Until the sixties of the 19th century, according to Levine, there were still 70 different time zones in America; around 1880 there were still 50. Each city had its own time (see Arte, 2018). The standardization was carried out by the railway companies and the meteorologists. In 1883, the railway companies set up four time zones for America, which were not passed by law until 1918. In addition, private companies such as the Observatory in Allegheny, the Horological Bureau at the Winchester Observatory in Yale or the Standard Time Company emerged, which promoted standardization efforts and for the first time offered "time" as a product by setting and testing watches or communicating the "correct" time (Levine, 1997 and 2016, p. 101ff.). Time clocks for the automatic determination of duration were developed in the 80s of the 19th century, which were sold almost monopolistically in the USA around 1907 by the International Recording Time Company (later IBM).

The standardisation of timekeeping was also socially consolidated by punctuality as a social status symbol or membership in so-called watch clubs (Levine, 1997 and 2016, p. 106ff.). This shows that, in the end, economic reasons were decisive in reducing the number of individual local times.

Global standardization was ultimately carried out in the form of world time. It is the result of a series of further developments: between 1884 and 1928, Greenwich Mean Time (GMT) applied, followed by Universal Time (UT) from 1928 and Coordinated Universal Time (UTC) from 1972. Their measurement is done today using atomic clocks and no longer in comparison with the Earth's rotation.

"The Coordinated World Time is the time standard from which the local times of all time zones are derived. The UTC is primarily based on the combined time measurement of hundreds of atomic clocks, expressed as International Atomic Time (TAI). [...] A day on Earth usually lasts slightly longer than 86,400 atomic seconds. On average, the deviation is about one millisecond." (Time, 2019).

The deviation of the mean solar time at the zero meridian, which is based on the actual Earth rotation, and the Coordinated World Time was exactly 37 seconds on February 5, 2019 (see Time, 2019).

c. Today's standardized characters

Technically, a common global syntax has emerged today. Even the leap second, leap year and leap day are synchronized worldwide. Thus, there is a globally uniform calendar and coordinated time zones. Even the division into a day into 24 hours, an hour into 60 minutes or a minute into 60 seconds is not questioned. The language is defined and agreed in ISO 8601:2004. It contains the characters used, including the separators for dates or time periods, the time of day, time zones, time periods and the scope of the year numbers. The standard allows only the years with the beginning of the Gregorian calendar (1583) up to the year 9999.

As a result, it can be stated that the standardization of time today has taken place globally and a decoupling from the natural phenomena of the sun and the moon has taken place, even if an individual will hardly feel these differences.

In addition, other globally standardized languages (signs of logistics) are pointed out here: Barcode, pallet formats, container formats, INCOTERMS, Global Trade Item Number (GTIN) by the organisation GS1, EDIFACT and their subsets, signs for mail traffic/Internet or the standards for voice transmission (telephone).

These technical solutions do not automatically overcome cultural hurdles. Because time management itself also enforces (external) pressure on all members of the interdependent system. In addition, there are also individual initiatives (self-compulsion through one's own goals) to control the entire system. "Hub firms" in supply networks can be cited as example.

The discipline and acceptance of external compulsion existing from a higher perspective can be expressed on the social level by a systemic higher conscience (Peter Kruse: self-control) or an accepted reason (Elias, 2017, p. 128). Thus, a common syntax still needs a common culture, i.e. a common understanding of values.

In view of the development of global value chains and the acceleration of innovation, the impression arises that all those involved in the global network have slipped into this new sense of time. This clarifies that the discussion on values must follow technical standardization. Read more about this in the following chapter.

3. Semantics: Time as an evaluative criterion
a. Quote: Everything has its time

This quotation comes from the preacher Solomon 3, 1-8 (see www.evangelisch.de/bibelstellen/prediger-31-8: "Everything has its time, and everything under heaven has its hour"). [Fn4]

Three interpretations and readings are possible for this quotation: (i) On the one hand, time is the basic representation of facts. And, **all** (!) facts are by definition related to time. I.e., **everything** has its time. (ii) On the other hand, everything has its right (!) time, i.e. not too early and not too late. I.e., **everything** has **its** time. (iii) Everything can also have a relation to a specific (!) superordinate time, i.e., everything has (only) **its** time.

Today's (daily) activities in society have 'their' time or 'their' duration. A few examples should illustrate this:

- Times at school/training: start of semester/school, training period, timetable, school year/ semester, examination periods, school leaving certificate, (semester) holidays,

- Times in working life: Working time/leisure time, bridge days/holiday days, working hours, periods of unemployment, sick leave/working days, start and end of work

- Times in sport: kick-off, playing time, break, stoppage time, final whistle

- Times in trade: Opening hours, shop closing hours, season start, sales, promotion days

- Government times: election day, polling station opening hours, electoral period, session duration, summer break, term of office, term of office or resignation

These examples could be continued, and they show that everything has its time. In the following, however, the second interpretation, i.e. everything has its time, takes centre stage: the temporal evaluation of individual or operational processes.

Before this explanation and discussion can take place, it must be clarified what is intended to be evaluated. This book focuses on business issues in the supply chain within and between companies, i.e. it looks at the individual processes whose holistic control is carried out in extreme cases from the raw materials to the end consumer. It can be assumed that the control (coordination) from the point of view of a purchasing company comprises the suppliers (Tier-1) and, if applicable, the suppliers of the suppliers (Tier-2). In individual industries, e.g. the automotive industry, longer control chains can also be identified.

For this reason, the basics of the processes are first set out before the discussion of the evaluations is then carried out.

b. Process as model of the real world

This section deals with the fact that all real observable issues can be represented as a process. In Figure 3.1, the elements and variants of processes are described and briefly explained as (a) to (d): (a) All circumstances have a beginning, an end and in between an executive activity with a process duration. This is to be called process element or process step. Thereby, the beginning and the end represent nodes of a process and the transmission to the next activity is called edge. (b) The element, which triggers the activities in the nodes and edges, is called order in the following. Such a representation is possible according to definition for all circumstances, i.e. for production, transport, storage, planning, analysis or control processes.

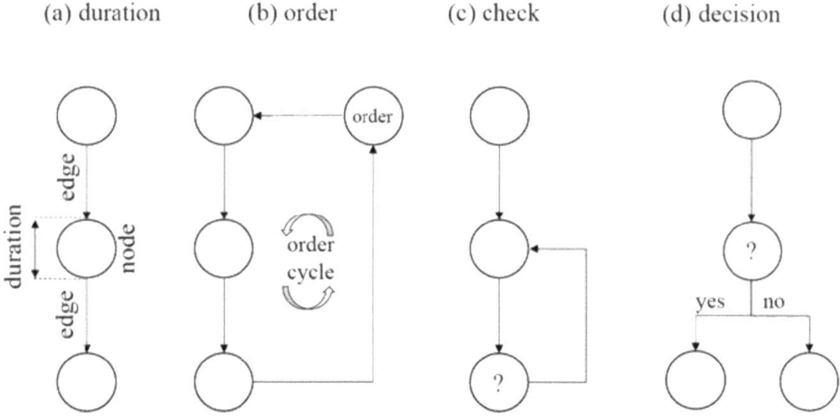

Figure 3.1: Elements and variants of processes

The entire process chain is referred to as the order cycle, and the time required is referred to as the lead time or delivery time. (c) If the contents of individual executions need to be checked, feedback may be necessary with regard to correctness. (d) If several alternatives exist, a decision must be made. The basic alternatives can, of course, also occur in a combination of variants (a) to (d).

In addition to the analysis of the individual activities, two other perspectives have a special meaning:

(i) The **hierarchy** of process observation: This means that a process is broken down to a detailed level (level 1) on an aggregated level (level 0). For example, a purchasing process (level 0) is the upper process for a number of sub-processes at level 1, e.g. determining purchasing requirements, selecting suppliers, agreeing on prices, concluding a sales contract, etc. The purchasing process (level 0) is the upper process for a number of sub-processes at level 1. This level 1 in turn can now be further broken down into level 2, where the process "determine demand" again consists of a number of individual processes, e.g. check inventory, determine production consumption, etc. This breaking down to a more detailed level can now be continued until the object of knowledge eludes an economic consideration. Figure 3.2 shows an example of a process hierarchy. Another way of hierarchizing processes is that each level is based on a very specific view. Thus, the organisation and its activities can be mapped on level 0, the data technical infrastructure on level 1, the physical-logistical infrastructure on level 2 and, for example, the mechanisms of coordination between the processes on level 3.

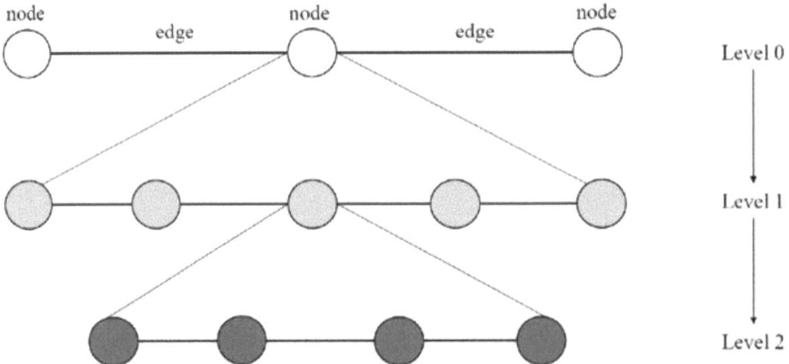

Figure 3.2: Hierarchy of processes

(ii) The **sequence** of processes: A single process has predecessor and successor processes. The examples of hierarchization have already illustrated this. Particularly in a highly labour-intensive and globally distributed structuring of value-added processes (outsourcing and offshoring), the structuring of the sequence of processes gains special significance. The sequence and the relations between the individual process elements then count to the foundations. Figure 3.3 shows the two basic sequences of processes.

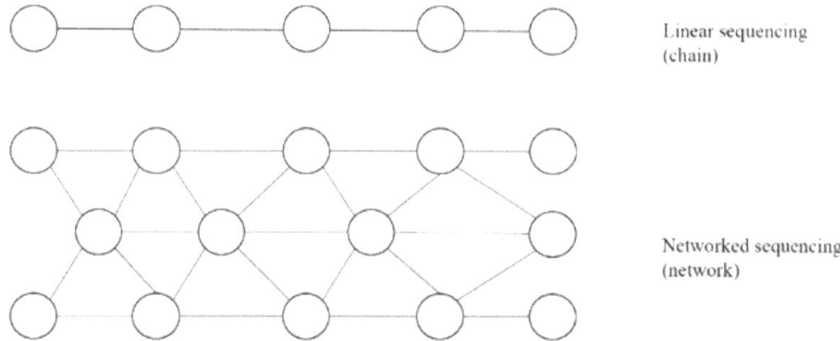

Figure 3.3: Sequences of processes

A purely linear sequence links the individual elements and progress is only possible after the predecessor process has been completed. Only one sequence is possible to reach the end of the process chain. The sequence of different process durations, the lack of timeliness or the lack of content quality can then lead to delays in the subsequent processes and then cause further additional processes, which assess the predecessor processes with regard to continuation reliability and, depending on these results, then make additional planning processes necessary. This is achieved with the aim of increasing the reaction speed and the process speed of the entire sequence. The total duration of the processes is determined by the length of the entire chain, the reliability of the execution and the sequence of the sequence program (if different). In a networked sequence of processes, there are other ways to choose alternative paths through the network to reach the end of the process or order fulfilment. This choice, in turn, depends on the already chosen paths of other process assignments and always requires planning and analysis elements to determine and control the program-dependent sequence and the quality of the execution of individual steps.

The spectrum of process sequences can then range from a linear sequence of a few process elements to a networked sequence of several elements. Figure 3.4 shows this using four variants of length and width. The more comprehensive term "network" is chosen for all forms of the process sequence.

If there are several possible paths through the network, the critical path is a very important and special path, i.e. a path in which there are no waiting times (buffer times) and a time delay of one element leads to a delay of the entire process chain. In this respect, the critical path should be known and should receive special attention.

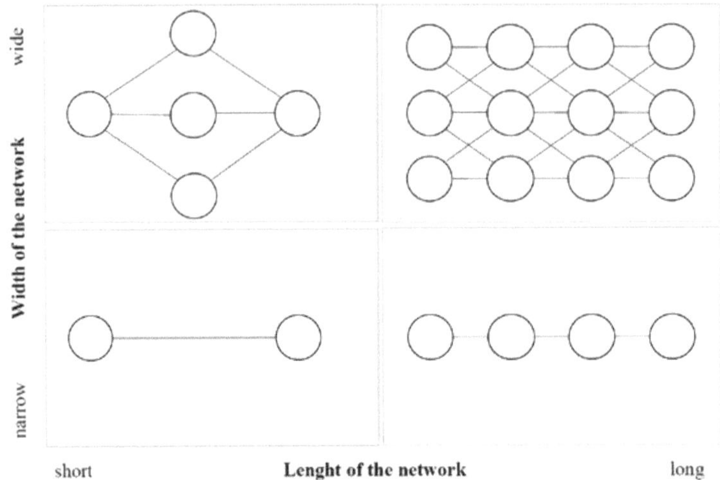

Figure 3.4: Length and width of networks

The possible time targets refer to the duration (lead time), reliability (punctuality), the capacity use of each step (usage time, idle time) and the waiting times of individual elements due to occupied capacities. Waiting times can be divided into capacity-based, sequence-based, planning-based and quality-based times.

The possible scheduling of the sequence can take place from the beginning of the process or from the end of the process chain. If a customer-oriented point of view is adopted, i.e. a customer has a time vision of his own processes and is looking for suitable suppliers with suitable preliminary processes, the starting point of the preliminary processes is determined within the framework of retrograde scheduling. The possibility of starting the pre-processes later is then regarded as a goal of process planning.

The sequence of the process assignment can be determined by an identical sequence or an individual sequence. The identical sequence is called a tact and can be found in the timetable for route trains, buses or trains. The individual sequences, on the other hand, are the result of planning that only applies to individual cases. Examples include scheduling for production in the workshop or the use of a taxi.

36

For the analysis and design of processes in a network, the three possible interdependencies must also be taken into account, as these make it necessary to coordinate processes among themselves:

- *Serial* interdependencies: The successive steps must also be coordinated with regard to the chronological sequence (see above) in order to avoid waiting times or supply chain interruptions.

- *Pooled* (bundled) interdependencies: The shared resources (nodes/edges) must be coordinated with each other with regard to their temporal availability and utilization in order to avoid waiting times or idle times.

- *Reciprocal* (mutual) interdependencies: The design of successive activities can only take place together, since each activity is determined by the activity of the other activity.

The scope of the analysis of processes thus depends on several influencing variables: Thus, the network is to be regarded as a system with its length and width, its relationship to neighbouring networks, the inner relationship in the network (interdependencies) and its performance claim (here: time). These parameters are listed in Table 3.1.

Parameters	Simple Analysis	Extended Analysis
Length of network	Short	Long
Width of network	Narrow	Wide
Embedding of the own network	Independent of surrounding networks	Dependent of surrounding networks
Time requirement	Reserves available	No reserves available
Interdependencies in the network	No sequence-, capacitance- or program-dependent relationships	Sequence-, capacitance- or program-dependent relationships

Table 3.1: Parameters influencing the process analysis

An excellent process flow (for all objects) is therefore a path through a network where there are no waiting times, no interruptions and no idle times: Everything happens in harmony. With identical times per job and per step, this situation can be achieved in the network. In the case of deviations per order or time requirements per process step, this situation is then no longer given, i.e. either the capacity in the node or in the edge waits for jobs or the job waits for free capacity.

The possible consequences of such a decrease in target achievement have been investigated and presented by Forrester and Goldratt/Cox. Since 1960, Forrester at the Massachusetts Institute of Technology (MIT) has been investigating dynamic processes in supply chains (here: in the sense of process chains) and determined that fluctuations occurring in a stable process sequence (i.e. a harmony) build up over the individual stages. He described this as a "bullwhip effect". Today, his didactic example can be found in the form of a four-stage process chain for the production and sale of beer and is therefore also referred to as a beer game. The causes of this build-up lie in the lack of coordination of activities in the process chain, since the informational basis of holistic coordination is not given, but each of the four participants seeks "his" optimum.

Goldratt/Cox (2013) have investigated bottlenecks in the production chain (here: process chain) and found that the throughput through a network is defined only by the constraint. This is the result of a program-dependent utilization of the capacity of a network node. The improvement of the throughput is therefore achieved by the coordinated program-related planning of this bottleneck. Their "Theory of Constraints" is also referred to as the "Drum Buffer Rope", since the bottleneck as a clock (drum) via a buffer (buffer) ensures the complete utilization of the process chain (rope).

In summary, it can be stated that the activities in and between companies can be mapped as a process in terms of time. The networks can be of different complexity and a unison is only possible with very specific parameters. The conflict of objectives between capacity utilization and processing time must be resolved by (the) management. The Bullwhip Effect and the Theory of Constraints have already scientifically addressed the basics of controlling processes in networks at an early stage.

After the presentation of the details of processes, the presentation and discussion of the evaluation of these processes now takes place with regard to three dimensions: duration or reliability, time as a reserve and time as a product.

c. Time as an expression of duration and reliability

A company's orders now pass through these networks (see Figure 3.4) and are evaluated over time. This allocation of a process run is made on the basis of the two/four criteria (i) speed/duration and (ii) reliability/punctuality.

The speed and duration of a process are two sides of the same coin. The physical speed is defined as distance travelled per unit of time. Thus, the duration is a direct expression of speed. As a result, the process is fast or slow in terms of duration.

The duration of the processes from the point of view of a company is the delivery time or the production throughput time. These two times have the same basic structure: In the "buy case", the delivery time corresponds to the lead time of the orders to the suppliers until the goods ordered have been received. For the "make case", on the other hand, the throughput of the company's own production orders until they are handed over to the delivery warehouse is the same. In both cases it concerns order-related throughput times.

The high speed and the short duration in a process chain are generally rated as positive: the fastest athlete wins the gold medal, or the fastest supplier receives the order. Speed also has direct advantages for value-added processes: a higher speed makes it possible to start the subsequent process later or to place the order later and thus newer information can be taken into account in the planning/control. Or a higher speed creates higher time buffers in the subsequent processes and thus either secures uncertainties better or enables a higher degree of temporal consolidation. In all cases, the higher speed has direct advantages and is therefore also assessed positively. Increased speed does not waste time, i.e. additional value-adding activities can be carried out.

The speed in an operational or inter-company process network can be increased in several ways:

- Fast execution of a process step (e.g. from ship transport to air freight)

- Faster start of a process step (e.g. due to available capacities or earlier availability of planning information within the scope of digitisation)

- Faster throughput through a production network through tight-pass control or better coordination (e.g. through shorter waiting times before the bottleneck)

- Faster throughput through parallelization of processes (e.g. through production synchronous production with just-in-time)

- Faster throughput by eliminating (unnecessary) process steps (e.g. by eliminating double quality checks)

On the other hand, an increased (in retrospect: excessive) speed, as can be observed for everyone in winter traffic, also contains risks. This increases the risk of accidents. Or by exceeding the maximum speed limit, a penalty is imposed in road traffic. Applied to operational processes, the quality of execution can also suffer if less care is taken and subsequent processes can therefore be negatively influenced. In addition, the employees involved, or the production equipment used can be exposed to increased stress/wear and thus impair the capacity available in the long term. All in all, those responsible for the accelerating and decelerating have to find the right measure to be successful in the short or long term.

The reliability of the execution of the order means the fulfilment of the agreed delivery time. Reliable execution is therefore always punctual. The target or agreed date must be specified with regard to its characteristics, i.e. the time window must be defined in terms of its duration. This can include a delivery day (e.g. 22 February), a delivery week (e.g. calendar week 44) or a special delivery time (e.g. before 8 a.m.). The statement on reliability can refer to an order or all orders. For example, 90 percent of deliveries are on time, 5 percent are 1 day late, and the remaining 5 percent were delivered two or more days late.

With reliability, the supplier provides a planning benefit for the receiver. Highly reliable performance requires no reliability inspections and no additional buffers to compensate for unreliability. On the other hand, unreliable processes either have to carry out these two balancing measures or they have to bear the consequences: interruptions in their own process chain or rescheduling to maintain it.

The speed or duration and the reliability or punctuality are a first dimension of operative time-based process evaluations. In the following, a second dimension of time assessment will be explained.

d. Time as a form of expression of a reserve

The utilisation or non-utilisation of capacities can also be expressed in terms of time. Capacity is an essential characteristic of a value network. It describes the flow rate in a node or edge. The incomplete utilization then leads to a reserve. This can take three different forms:

(1) Capacity reserve as unused time: This is the possibility of being able to provide free capacity for new orders at short notice and not to delay the progress of the

order due to waiting times due to capacity. In this way, the timely allocation of free capacity creates a benefit that does not result from the direct evaluation of the individual order, but from the evaluation of the order program (program-related benefit). Time links the individual program elements and makes coordinated planning and control necessary.

(2)   Stocks as a temporary buffer: This is the possibility of being able to deliver temporal order fluctuations at short notice due to stocks that have already been produced. Thus, foresighted or forecast-based planning creates a temporal value for the next stage.

(3)   Reserve in form of possible waiting times: This means that a follow-up process does not have to be executed immediately and a standstill does not jeopardize order fulfilment. The costs of waiting must be weighed against the chances of waiting. For example, time consolidation (e.g. transport consolidation with the goal of fully utilized means of transport) can make the total of all orders more economical. Here other, already waiting orders provide a benefit for the later arriving orders.

These characteristics implicitly assume a conflict of objectives between the capacity costs and the service benefit over time. Higher capacity utilization (and thus lower proportionate idle capacity costs than cost unit overheads) is offset by longer processing times and thus a lower service benefit of the orders (cost units) over time.

Goldratt/Cox have resolved this conflicting view to some extent with their Theory of Constraints. With a limited capacity and consequently longer throughput times, they have dealt with the possibility of better bottleneck control. With their bottleneck-related planning, they design a way of increasing throughput, i.e. shorter waiting times and thus a shorter throughput time, while at the same time achieving high capacity utilisation. This view is the consideration of time as an expression of a capacity reserve through better planning, i.e. an unused potential. Their approach does not resolve bottlenecks in principle but seeks to achieve a minimum waiting time for orders when capacity is fully utilized.

e. Time as a unique product

In addition to duration/reliability and reserve, time can also be a relevant criterion for operational activities/processes as a unique product. Time with regard to this effect can have several characteristics.

Time can have a strategic relevance ("active duration") as the duration within which a new and for the first time area-wide competition-decisive technology is to be introduced. In this case, the aim is to achieve a short time-to-market. One example is electromobility, in which those responsible at car manufacturers announce that they will be offering new e-cars up to a certain point in time in order to further extend the lead over traditional competitors and thus their own strong market position.

In addition, the period of time during which a comparable offer is to be on the market in order to prevent a competitive superiority (monopoly) ("reactive period") can be regarded as a strategic criterion. This is the subject of the interview with Morozov, who describes the superiority of Google, Amazon and Facebook and complains about the lack of European providers.

Thirdly, time can be classified as a strategic product, i.e. a product is offered that provides the user additional time for alternative uses. For example, the ICE or the Concorde can be mentioned, which allow a shorter travel time. But also, domestic helpers are included to have more time for "themselves". Free time can then be used to create an additional benefit; the faster process can of course also have an emotional benefit.

In the area of infrastructure, these include rescue services, the technical relief organisation, emergency call facilities, fire brigades, storage service providers for blood preserves, energy and medicines. Their service is to provide rapid assistance in the event of a crisis. They initiate the restoration of the readiness to perform within short time with it.

In addition, time can be regarded as an essential product feature, i.e. the service package can be compared in terms of several features and is competitive in terms of time benefits. As an example, mail-order/distance selling orders with 'Same Day Delivery' or messaging services that can exchange 'real-time' messages can be mentioned. In addition to social media, providers such as riskmethods can also be mentioned here, which display a critical time status in the supply chain in real time.

As the last characteristic, time in the production area is regarded as the essential product characteristic, e.g. within the framework of just-in-time replenishment control: In addition to the synchronous production of manufacturers and suppliers, this also takes place via the short-term retrieval of individual components and thus the assurance of short-term individual market management.

These characteristics justify the aspects of "time" as a strategically relevant criterion.

f. Industry 4.0 and the time-based mapping of business processes

Digitalization takes up the value of time for flexible and customer-specific planning or control and the 4.0 concepts are discussed as the necessary and success-securing future formula. This topic dominates the agenda at congresses and in professional journals. The impact of these developments is strongly influenced by the factor "time". In the foreground of the public discussions are the technical aspects, which make such a change possible in the first place:

- Internet of Things (IoT) as an expression of the manageability of all system elements in the network,

- a widely used sensor system for recording the status of machines, materials and environmental conditions,

- the cyber-physical systems (CPS) for the automatic transmission and networking of the systems in a network with independent self-control and

- the Smart Factory as the final expression of self-controlling decentralized manufacturing processes in the value chain.

Obermaier provides this essentially technical definition of industry 4.0 with an organizational business reference. ""Industry 4.0" is a form of industrial value creation that is characterized by digitisation, automation and networking of all actors involved in value creation and affects processes, products and business models of industrial companies." (Obermaier, 2016, p. 8). It thus combines the technical with the business aspects.

Through the integration of cyber-physical systems in production and logistics, the individual sub-steps of production can now be networked without media disruptions. All system states are permanently recorded and made available to all relevant subsequent stages in real time. Thus, all processes and states are technically transparent and can be used for production planning and control. Another important feature is the decentralized decision-making capability of each system element. For this purpose, the value chains are networked horizontally (i.e. all stages of the value chain) and vertically (i.e. in all hierarchical stages of production planning and control). As a result, cooperation in the supply chain changes fundamentally:

"Value creation no longer takes place sequentially and with a time lag, but in a network of constantly communicating and flexibly reacting units that largely organize themselves." (Roland Berger, 2015, p. 17).

According to Obermaier, the effects are expressed in terms of improved products, improved processes and improved business models. The basic objectives of digitisation are competitiveness and massive cost savings (see e.g. Schuh et al., 2017). The simultaneity of events and data availability and the timeliness of relevant facts should ensure the future of competitiveness. Immediate availability plays a central role as an assessment standard in competition. With the help of digitisation, the following fundamental effects are aimed for:

- Initial data input as a prerequisite for generating additional information

- Fast data input as a prerequisite for generating additional information

- Fast result output through timely analysis and automated decisions

- New kind of result output in terms of first-time available findings, decisions and services

Digitalisation will then be discussed in the concrete implementation on the basis of two basic characteristics:

(1) Digitalisation in the sense of automation of an activity, i.e. from manual execution to automated execution. Previously manual activities are prepared by software algorithms and carried out in the form of automated or partially automated production systems. As a result, the products can be manufactured faster/more precisely.

(2) Digitalisation in the sense of continuous networking, i.e. available data are used for the purpose of greater transparency ("descriptive"), better predictions ("predictive") and the establishment of self-control ("prescriptive"). For this purpose, the data basis for data acquisition must be expanded or built up and data processing for the purpose of information acquisition (data evaluation, decision preparation or automated execution) must be carried out by software. Thus, the available data are used for the first time, processed faster, the

44

decisions better founded and made faster. In previous work I have described this cascade of data use as information products 1, 2, 3 or 4 and the dividing line between the digital and analogue worlds as a "digital penetration point" (see Darr, 2017b, p. 57ff. and 2019, p. 34ff. and p. 47ff.). [Fn5]

The factor "time" is decisive for the high expected success of digital transformation. This means that more data is collected that would never be collected in the analogue world for economic reasons, and it is processed more quickly into information. According to Obermaier (2016), this better information can then be transformed into improved products (e.g. streaming), improved processes (e.g. better resilience, tailor-made steel or condition monitoring) or new business models (e.g. networked dairy farming). Or, according to Darr (2017a), the better information can be expressed in more secure processes (e.g. real-time tracking and better control), higher economic efficiency (e.g. through better risk control) or better sales market opportunities (e.g. by comparing customer wishes with performance).

g. Brief interim conclusion

In summary, it can be stated that "time" is relevant as an evaluative criterion for operational and strategic issues and will gain relevance through digital developments. The links in the value chain require a universal criterion for comparative assessments for coordination: "time" is thus the accepted global language and the basis of the evaluations.

The two chapters on syntax and semantics provide the prerequisites for the third stage of semiotics, pragmatics. Together with standardization and evaluation criteria, they form the foundation of meaningful and purposeful organization of time management in companies. This is done first for the operational processes and then for strategies. Finally, these statements are embedded in the societal discussion of time, which is also discussed in terms of acceleration.

4. Pragmatics: The paradigm in operational management
a. Quote: Time is money (Zeit ist Geld)

This worldwide known quote comes from Benjamin Franklin in 1748 from his book 'Advice to a young tradesman': "Remember that Time is Money". The basic idea of Franklin, which he expresses with his statement, defines the claim of efficiency by avoiding waste, here the waste of time by useless sitting around for half a day: "Remember that TIME is Money. He that can earn Ten Shillings a Day by his Labour, and goes abroad, or sits idle one half of that Day [...]." (Franklin, 1748).

This ideal of increasing efficiency by avoiding waste has also been used several times outside business administration: Michael Ende, for example, writes in his book Momo in Chapter 6 'The calculation is wrong and yet works':

"Because time is life. And life resides in the heart. And nobody knew that better than the grey gentlemen. No one knew the value of an hour, a minute, even a single second of life as they did." (Ende, 2005, p. 61).

The waste of time is expressed by the different time spent by the hairdresser Fusi, e.g. the marriage of Miss Daria, a quarter of an hour in the evening sitting at the window, avoiding time-consuming conversations with customers, by the grey gentlemen in the high sum of wasted seconds throughout the life of the hairdresser Fusi. It is an extreme and exaggerated version of the "idle time" in Franklin's quote.

"No one seemed to realize that by saving time, he was actually saving something quite different. No one wanted to admit that his life was getting poorer, more uniform and colder." (Ende, 2005, p. 78).

Michael Ende thus makes it clear in which extreme efficiency can be pronounced and that the time saved cannot be the only goal in life. Franklin does not go to the extremes of the grey gentlemen in his advice, but from his point of view he also subordinates his life to work. Another current example of the conflicting goals of efficient use of time can be found in an article in the Frankfurter Allgemeine from 22.02.2019: "Delivery until yesterday, please. [...] A wholesaler [...] is bankrupt." The high demand for service in connection with the insolvency of the supplier, however, draws attention to a tense relationship in the implementation of an efficiency goal.

Charly Chaplin also illustrated the dark side of this efficiency in his film *Modern Times*: two screws had to be tightened on an assembly line station. In spite of the too high pace, he tries to do his job and gets into the machine. Later, he continues this step of "tightening the screws" in a surreal and comical way in everyday situations. This corresponds to the comedy and the message of the film. The alienation of the work through its excessive efficiency is thus not only a theme in Michael Ende's work, but also in Charlie Chaplin's.

The film "In Time - Deine Zeit läuft ab" is also dedicated to this topic with the same message. There is a worldwide economic system with only one currency: lifetime. Everyone has a maximum of 25 years plus one year. All consumer goods, e.g. a cup of coffee or rent, have "time prices". Time can also be given away, e.g. to the rich, to time beggars or for lunch. The scarce factor "time" requires a conscious approach in its use, i.e. efficient action becomes necessary. Excessive consumption means an earlier death. The film draws its tension from the scarcity of the resource "time" and the possibility that this can also be stored in time capsules and then means infinite life. In addition, there is the game of the crooks ("Minute Men") with the time police ("Game Keeper").

Efficient organizations are nothing new in today's business administration; on the contrary, increasing efficiency is an expression of rational behavior and an essential goal of management. For example, Ford and Taylor made this efficiency the central pillar of their considerations about 100 years ago: Frederick Taylor has used the term "Scientific Management" to study work processes by breaking down the entire activities into several monotonous repetitive sections and determining standard times. These could then be used as planning basis (Taylor, 1919). Especially in logistics, project management and the digitisation of processes, this knowledge also plays a central role today. Ford, on the other hand, split the production steps of the T-model into timed sections in order to increase efficiency by means of assembly line production.

The importance of efficiency can be seen from the use of time management in journals: An analysis of the term 'time management' in the databases of Emerald Journal (www.emeraldinsight.com) or EBSCO shows that no entries were found for the years 1930 to 1970. In the 1980s, there were individual indications of 'shopping behaviour'. At the beginning of the 1990s, Just-in-Time was increasingly discussed intensively. Especially from the year 2010, the number of mentions rises. On 09 February 2019, 227,827 entries were made in the Emerald Journal.

An essential pillar of the time discussion stems from the Toyota Production System, which serves as a worldwide concept for increasing efficiency by avoiding waste. Individual measures are worth mentioning: Just-in-time deliveries, Kanban replenishment controls, Kaizen as a learning platform for continuous improvement, continuous material flow, standardization and process timing.

The need to improve or increase efficiency is still omnipresent and undisputed today. Since the 1990s, the high degree of division of labour between different activities has been reflected in a high degree of (global) division of labour between companies in the supply chain (outsourcing and offshoring). In this respect, the increase in the number of research names from the 1990s onwards is not surprising. In all cases, the time saved is money.

b. Principle of valuation

Strictly speaking, time is not a scarce commodity, but is used as an indicator of efficiency or inefficiency: the use of production factors (e.g. employee working time, machine running time, storage/waiting time) is measured over time and assessed in terms of its effect on the output (e.g. quality, delivery time, punctuality, time of market entry) of the work step. The time required is thus an indicator of the costs or performance of a product.

The evaluation of efficiency or inefficiency is basically based on the analysis of the temporal use/waste of resources in the supply chain or production. All activities are classified as 'value adding' and 'non value adding' in terms of their contribution to the final product. Non-value-adding' activities include those activities which, by omitting them, have no influence on the assessment of the product by the customer. This analysis takes place in several ways.

In order to clarify the principle of process evaluation in relation to time, a well-known group exercise is briefly described which illustrates the coordination of a process on the basis of a time objective: There are eight participants in a circle (see Figure 4.1) who have the task of passing through a ball once with each participant. The order of the run and the organisation can be decided freely by the group. In the start variant, such a run may look like the one described by "slow circulation round" in the left part of Figure 4.1. One participant starts, considers briefly and throws the ball to another participant. This process is repeated until the last participant has received the ball. A moderator measures the entire process time. He then asks the group to complete the process in a shorter time.

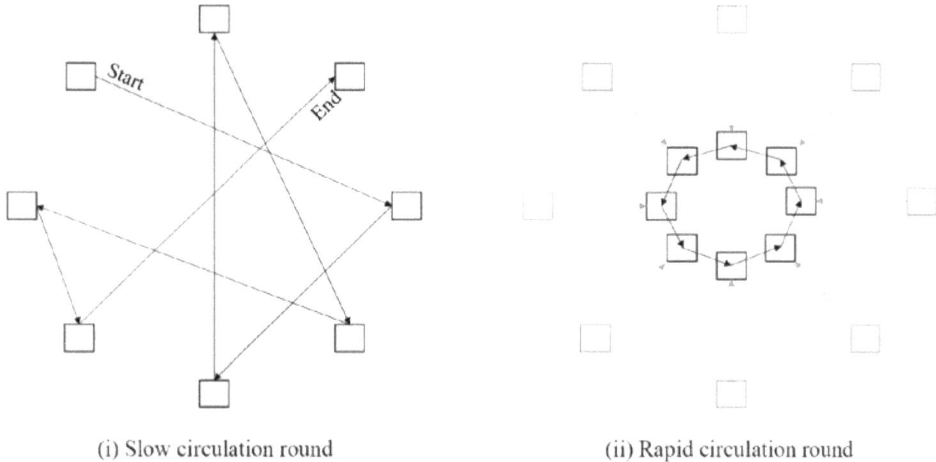

Figure 4.1: Slow and rapid circulation

Now the group starts to run the process faster, i.e. more efficiently, by better organization (e.g. distance between the participants and decision on the order of the run). This learning process in such a group exercise can take place in several runs. It could then look like the right part of Figure 4.1 ("rapid circulation round"): the participants stand close to each other and the passing on is not reconsidered every time, but the ball is always passed on to the right neighbour. As a result, time is saved due to the shorter distance between the participants (i.e. organisation) and the time saved for reflection and decision making (i.e. coordination). In self-made experiences of this exercise, an excellent total time of approx. 1 to 2 seconds is achieved with approx. 15 participants.

The principle of assessing operational management activities in this book is based on looking at them in a sequence, i.e. as a process. It does not only consider an isolated activity step, but the sequence of several individual subprocesses and their activities. This process-related view is the view assumed here and is suitable for achieving the time targets. It is supported by the high division of labour in the (global) supply chain, which usually begins with an order and ends with the fulfilment of the customer's wishes. For the sake of completeness, it should be noted that efficiency can refer not only to the entire process, but also to an isolated activity. The origins of the procedural view in business administration go back to Nordsieck (1934) and his workflow organization.

The organisation of all activities (here: operative activities) is ideal/perfect when everything "flows", i.e. all processes move in unison, there is no/very little distance to be covered between the activities and, despite all the chains, the customer's wishes are always immediately fulfilled without there being over- or underproduction. This ideal is unfortunately an operational utopia in view of the uncertainty and lack of predictability of demand, the fluctuations over time, the transport distances and the permanent changes in framework conditions. Digitisation is also unlikely to overcome these hurdles in the foreseeable future.

Therefore, a management approach is needed that achieves 'good' results in the face of the challenges: Thus, the direction of a company has to prepare the following nine structural decisions, so that the processes can be executed in the sense of a quick run: (i) the definition of strategies (customer benefit profiles and their fulfillment by business processes), (ii) the relevant objectives, here: time goals and their role in the enterprise, (iii) the classification of the enterprise into isolable subsystems, e.g. also to suppliers or customers, (iv) the hierarchisation of entrepreneurial areas and their responsibilities, (v) the definition of evaluation standards in relation to the strategies, (vi) the organization of the coordination of conflicts of interest (rationality), (vii) the definition of stable (efficient) decision-making chains, (viii) the further development of existing structures and processes (innovation) and (ix) the control of previous executions and the resulting learning processes. Only within this structural framework do the operational and inter-company processes take place.

In summary, it can be said that the temporal use of resources is embedded in a series of structural preliminary decisions of a company. The business processes themselves are then assessed with regard to their use and their effect on the end product. Time is the valuable asset here and must not be wasted or used imprudently, i.e. the use of resources and the organisation of operational processes must be carried out according to time efficiency criteria. Their waste reduces efficiency, increases full unit costs, lowers the benefit endowment and ultimately lowers the contribution margin for the offering company. This raises the question of which methods to use to achieve this.

c. Methods for analysing and controlling time

Several methods have been established for the analysis and control of processes under time aspects in business administration:

(1) In **value analysis**, this is done with regard to the constructive characteristics of a product. The installation of 'unnecessary' components leads to the waste of resources. Likewise, the choice of unfavourable production speeds of machines can increase the consumption of resources or reduce the quality of the products. Time studies (see Taylor) can also reveal inefficient work processes of employees (e.g. Schönsleben, 2011).

(2) The **Total Cost of Ownership** approach considers a holistic evaluation of temporal processes, in particular the effects on other departments/processes. Time can also be an indicator of opportunity costs if they occur later in the supply chain or with a time lag: fast production processes may cause quality defects that later lead to complaints or interruptions in production.

(3) The analysis can also begin with the seven types of waste developed by Taiichi Ohno (2013) in **Lean Management**: First, the section or object of possible waste is narrowed down: transport, stocks, movement, waiting, overproduction, technology/ processes or rework by rejects. Even if the individual types of waste are not free of overlap, for Taiichi Ohno they were the right way to track down the 'evil', i.e. the detrimental effect (the output) of starting the analysis. The analyst's creativity then lies in the decomposition of the overall process and the associated evaluation of each section of the seven types of waste. Possible overlapping characteristics, e.g. higher stocks due to longer waiting times, must then be neutralized by the analyst.

(4) The methods of **process analysis** and value stream analysis are fundamentally different. They begin with the input of the processes: the process analysis subdivides the individual processes into sub-processes or activity steps and its components in order to gain an insight into the weak points and thus for the inefficiency. The creativity of the analyst lies here in the determination of disadvantageous facts, e.g. no clear description of the activities of the process, uncoordinated interfaces and/or media breaks, redundant activities, no specific regulations and/or no generally valid regulations or no and/or unclear responsibilities.

Inefficient solutions in the production cycle can lead to interruptions in production time or to traffic jams in front of machines (waiting times), which are thus classified as 'not adding value'. The effects of an interdependent production program can also lead to waste, as the lack of coordination of batch sizes and processing times can lead to waiting times for products or idle times for production equipment. In both cases, time (waiting time or free time capacity) is the indicator of inefficiency.

The best method of process management or process design is based on the nine decisions to the infrastructure and the application of the following principles to obtain an efficient run:

- *Holistic*, i.e. consideration of the entire process chain, definition of the respective responsibilities, transparency of steps and states and knowledge of the process performance(s)

- *Process design*, i.e. description of the individual tasks (contents with the starting point, the activity and the end point), the variants or branches, the linking of the individual process steps and their conflicts of objectives

- *Result orientation*, i.e. definition of output/result criteria and their measurement, regulations on sustainability or robustness and criteria on the process competence of the parties involved

- *Control rules*, i.e. possibilities to standardize, eliminate, outsource, pool, parallelize, re-segment or relocate

This process management design has been discussed extensively in the literature for interested readers, e.g. by Armistead (1996), Becker (2008), Bögel et al. (2014), Burlton (2001), Kim/Ramkaran (2004), Klepzig/Schmidt (1997), Kotzab/Otto (2004), Liebetruth (2016), Ohno (2013) or vom Brocke et al. (2014).

(5) **Value stream mapping** is a business method for identifying non-value-adding activities/processes. By redesigning the processes, especially the non-value-adding activities, an improved value stream is to be made possible.

At the core of this analysis, the share of the pure processing time in the total processing time is determined. The basic hypothesis is that the pure machining time is only a fraction ($< 5\ \%$) of the total lead time. The main times are non-value-adding resting/waiting times. By redesigning the design or the sequence of the process, these waiting times are then to be minimized or avoided.

The value stream analysis proves to be very practice- and implementation-friendly by the possible use of the representation of the sequences even with paper and pencil. It enables simple yet clear communication with all parties involved in the process. For networked production processes, on the other hand, a visual representation can quickly appear confusing. The complete process chain usually includes an order cycle from the customer order through individual production steps to delivery. Standardized symbols are also used for the individual process steps and activities/activities: Customer/supplier, process step, work station and data box. In the latter, the processing time ("cycle time") and the processing times are recorded.

The proposals to increase the share of value added in the total lead time largely coincide with those of process analysis.

(6) The analysis of the time sequences, irrespective of the choice of method, provides companies with temporal and thus economic advantages. **Stalk/Hout** (1990) discussed the question early on. As a starting point they saw the lower temporal productivity of enterprises. They summarized their statistical findings in the form of four "Rules of Response" (Stalk/Hout, 1990, p. 76ff.):

- The "*0.05 to 5 Rule*": Here the authors confirm that the productive time is between 0.05 and 5 percent of the total lead time.

- The "*3/3 Rule*": Here they confirm the three main reasons for the poor productivity in terms of time: (i) The processing time of an individual order is linked via interdependencies with that of other orders, (ii) plans for the sequence of orders are frequently revised and (iii) individual new prioritisations of the order sequence are made at short notice by the management.

- The "*¼-2-20 Rule*": For any reduction in order size to 25%, labour productivity or capital employed productivity can be doubled and costs reduced by 20%.

- The "*3x2 Rule*": By improving productivity over time, a threefold increase in turnover and a doubling of profitability compared to the competition can be achieved: "Time-based competitors outperform their industry" (Stalk/Hout, 1990, p. 2 and the numerous examples from Chapter 1).

The two authors see the starting points in a process-oriented organization of the company. They have also outlined the process chain on the basis of throughput times and the proportion of value added over time and summarized it in figures on pages 66 to 76 similar to the value stream analysis. The procedures for achieving these enormous benefits are based on defined cornerstones (Stalk/Hout, 1990, pp. 200-209). The production processes can be organized on the basis of their handling in such a way that the planning principles can convert a function-specialized enterprise into a process-oriented enterprise:

i. The business processes of the company are represented in a diagram ("mapping").

ii. The central aspects include those involved in the process, the interfaces and the need for coordination.

iii. The critical path through the company is known and is actively controlled.

iv. The business process model is deliberately kept simple and comprises only the essential processes and decisions.

v. Functional views of individual areas/departments must be subordinated to the process organization of the company.

vi. Key figures at "time" (in contrast to costs) serve as primary control indicators.

In addition to the analysis tools described above (process analysis, value stream analysis), the authors also make concrete suggestions for the design and management objectives of the content.

(7) The operational throughput of production orders is dealt with within the framework of **scheduling** and is gaining in importance, particularly in series production within the framework of workshop, flow or group production. Mathematic methods of optimization fail due to the complexity of the task due to the effectiveness and solution defects. In this respect, heuristics are used as auxiliary algorithms ("priority rules"), e.g. the "Shortest remaining processing time rule", the "Shortest operating time rule" or the "Delivery date rule". These procedures lead to feasible sequences of the sequence program within a manageable time frame.

(8) Nyhuis et al. (2006) use the funnel model to examine the operational processes and thus represent the incoming, the waiting and the completed orders. They evaluate the analysis of the conditions in production or the warehouse on the basis of the **"logistics curves"**. The individual services (receipt, processing, dispatch) are graphically illustrated over the course of time and the stocks or the ranges per analysis object (article) are shown. Using "ideal curves", inventory reductions can then be achieved without service losses, for example.

(9) Harrison/van Hoek are developing a concept for **"supply pipeline performance"** for time-related (process-related) analysis and control (Harrison/van Hoek, 2008, p. 148ff.): They distinguish between P-time and D-time. P-time (production) records the processing time of an order by the company. The D-time (distribution), on the other hand, reflects the time expected by the customer to the delivery time of the company. If the D-time is shorter than the P-time, management needs to perform time analyses to shorten the P-time. Graphical measures also clarify

the time requirements of the value creation or non-value creation of the individual activities and then suggest a changed organization of the processes. These range from improving process know-how, reducing product diversity or batch size, improving internal coordination between departments, shifting order penetration points (OPP) or automation to reduce production time (Harrison/ van Hoek, 2008, p. 163ff.).

In summary, it can be said that the individual methods for analysis and control have different key points. The basic idea behind all proposals is identical: Planning and control are dominated by shorter turnaround times.

d. Time in purchasing, production and distribution

The lead time of business processes through a company is manifold: Basically, the order is the triggering impulse, and it can be regarded as the triggering of a sequence of activities (Darr, 1992, p. 18). Orders come primarily from the customers of the enterprise (customer orders). The throughput now depends on the design of the order penetration points and varies from pure delivery orders from the distribution warehouse through assemble-to-order orders to make-to-order orders. In the "make case", suborders can now be placed from production for the distribution warehouse or parts warehouse. In the "buy case", on the other hand, procurement orders are to be directed to the suppliers.

Sales or distribution processes do not only take place in distance selling under strict observance of the lead time. It is emphasized that the principles of process analysis (see above) should be applied here to ensure interfaces and delivery time or delivery reliability.

Also, in the purchase temporal processes are important, because a shortened procurement time make additional temporal room for maneuver in the own manufacturing possible (here: from view of the entire customer-referred throughput time) or open additional possibilities at the sales market by fast and flexible goods supplies of the suppliers.

Purchasers have to analyse and design the process chain with suppliers as well as the production chains, even if the development stage of a process-oriented purchasing management is still in its infancy. Not the pure superficial cost consideration, but the time consideration with intended cost advantages should be the requirement of a process management in purchasing. The same principles and findings apply here as in Stalk/Hout (1990).

e. Key question of operational planning: accelerated or decelerated

The previous remarks have confirmed the core statement that "time is money" or "faster also means more efficient". The statements of Stalk/Hout or Harrison/van Hoek, the process analysis or the value stream analysis speak for themselves. This uniform interpretation of higher speed is fundamentally questioned by Delfmann (2010) by raising the question of whether deceleration, decoupling or deconsolidation should not lead to a reassessment (i.e. a paradigmatic change of perspective) of procedural decisions. He thus poses the fundamental question of whether acceleration, coupling and consolidation of processes and supply chain structures are really the right strategy. Figure 4.2 expresses these two paradigmatic poles by curve 1 (acceleration) and curve 2 (deceleration). For reasons of simplification, the measurement criterion of advantageousness is called "success". In the case of "acceleration" there is only the claim of permanent temporal improvement (optimum "1"), whereas in the case of "deceleration" optimum "2" has to be determined.

Delfmann's arguments for deceleration are as follows: the time delay of processes brings economic advantages. He cites the temporal transport consolidation, the increase in safety stocks in warehouses or the increase in batch size as examples. The disadvantage of longer delivery times is offset by lower average unit costs and more robust process chains. The management has to weigh the opposing evaluations against each other, i.e. to find an optimum time "2".

Decoupled networks (structures with less complexity) lead to less coordination effort, fewer program effects in the process planning and are therefore easier and more robust to control. Deconsolidated networks, i.e. the low degree of centralisation of coordination and the low degree of centralisation of the configuration of supply chains, will be less susceptible to environmental changes or crises (robustness) and will be able to restore their performance more easily in the event of a crisis (resilience). Overall, the reliability of such networks increases.

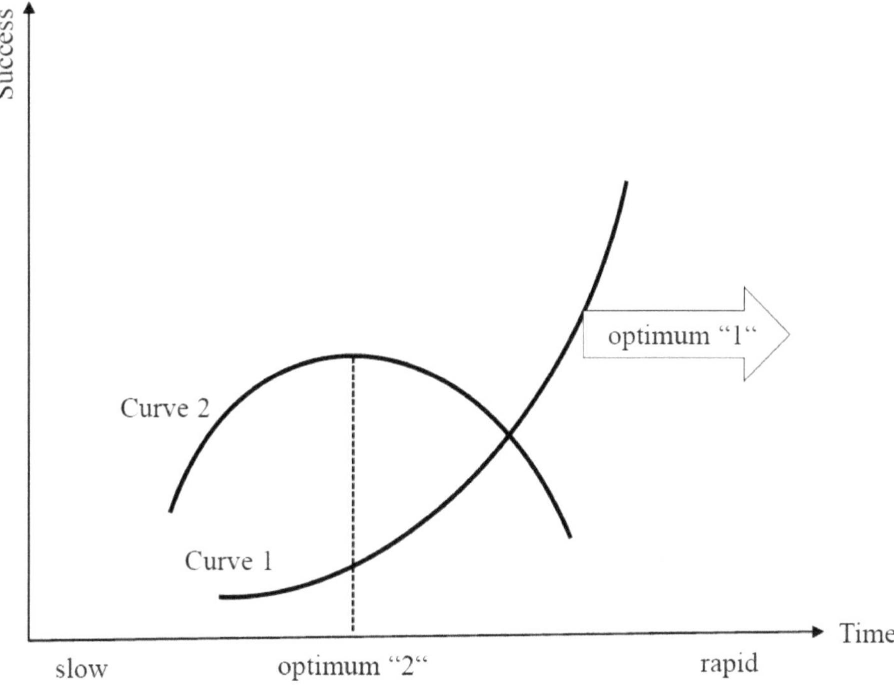

Figure 4.2: Acceleration or deceleration

Thus, Delfmann favours an integrative consideration of the advantages of acceleration and deceleration and a balancing of these opposing effects. He thus rejects the one-sided unchecked temporal shortening of all processes.

f. The evaluation of operational processes against the background of the discussion of industry 4.0

The key concept of the digitised supply chain in the form of an industry 4.0 has already been briefly explained in Chapter 3. Here the factor "time" is significantly influenced:

-   The widespread use of sensors makes it possible to greatly expand the amount of data available. On the one hand, this more extensive collection process can be designed economically, and, on the other hand, this data can be collected and transmitted relatively quickly, i.e. without long delays. In the context of risk management, for example, software applications (e.g. from riskmethods) can be used to inform the buyer in real time about disruptions in the supply chain. This enables faster conclusions to be drawn in the sense of faster measures.

- The cyber-physical systems allow these 'real-time' transmitted data to be used to adapt the production control in order to avoid interruptions or to make the production processes more efficient.

- The digitized purchasing applications in state 2 or 3 (information product 2 or 3, see Darr, 2017a, p. 58ff.) allow changes to the cost or risk assessment of procurement orders to be made available to decision-makers in a timely manner. Thus, more uneconomical decisions can then be recognized more quickly (or at all).

- The intelligent applications, i.e. the data-processing software applications, can determine and point out critical situations faster due to their processing speed or judge decision variants regarding their advantageousness faster.

The digital possibilities, which are reflected in higher information products and ultimately drastically reduce waiting times (see Stalk/Hout: The 0.05 to 5 Rule), can be illustrated by Figure 4.3.

The starting point in the analogue world (state 0, see Darr, 2019, p. 30-37) is point A with a process time Z_1 and process costs K_1. An accelerated process time (from Z_1 to Z_2) results in an increase in costs from K_1 to K_2 (point B). The cost curve thereby represents the cost efficiency line regarding the time. By switching in the digital world from state 0 to state 2 or 3, i.e. setting up machine-readable data along the process chain of an order cycle, processing data to information by software applications, automated processing of data in the sense of analyses (state 2) or decision proposals (state 3) and avoiding media breaks, waiting times in the throughput can be significantly reduced and the process time significantly shortened (point C). This can even be done at lower costs ($\Delta K = K_1 - K_0$) than in the analogue world. In the digital world, decision-makers can even achieve an even shorter process time Z_3 ($\Delta Z = Z_2 - Z_3$ in point D) at the same cost level K_1 of the analogue world (based on the cost curves of Figure 4.3).

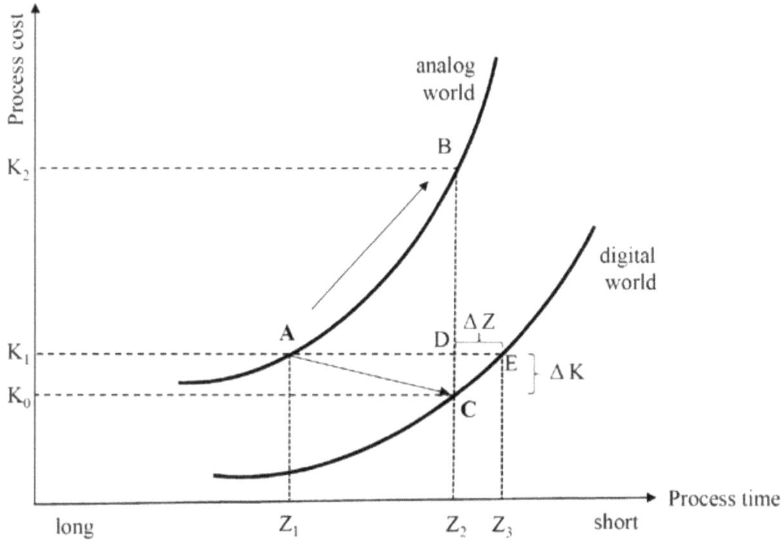

Figure 4.3: Effects of digitisation and digitalisation

The economic effects of digital process solutions should not be limited to the considered process alone. This process concentrates on the automated overcoming of interfaces and media breaks and the automated evaluation of facts (state 2) or decision alternatives (state 3) and leads to a flatter cost curve in the digital world according to Figure 4.3. A holistic process view, which is represented by the order cycle with a defined delivery time in Figure 4.4, therefore includes the interaction of the data processes with the material processes. Thus, the shorter process time Z_2 can lead to the following two fundamental reorganisations of the material processes (see Darr, 1992, p. 320ff. and p. 346ff.):

- Reorganization of the material processes (and their underlying structures) at the same delivery time level, i.e. shorter time for data processes and planning of a longer time for material processes. Here the temporal shortenings are used for longer production times, which means the overall throughput time remains unchanged and the benefits are reflected in lower production costs.

- Use of the reduced process time for profiling on the customer market; see chapter 5. Here the time reductions are used directly for better service offerings for the customers and the advantages are expressed in higher sales successes.

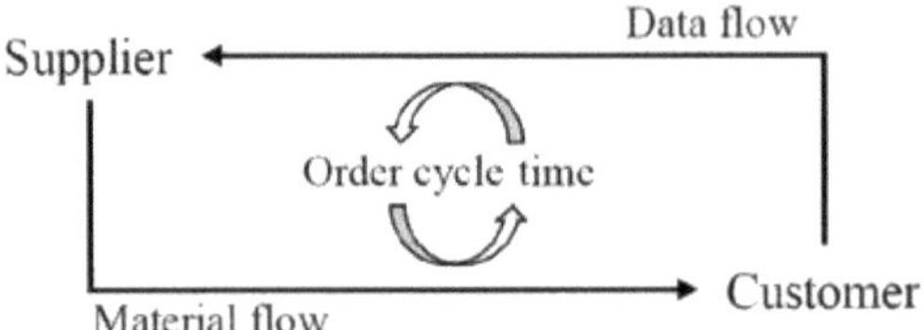

Figure 4.4: Order cycle and time effects

The opportunities offered by digitalization, which basically lie in a foundation of benefits in terms of process security, securing results or securing uniqueness for the customer, initially lead to an operational acceleration of data processes in order to prepare and make decisions more flexible and quicker in the supply chain.

g. Interim conclusion

Time plays an important role in the operational management of a supply chain, since process reliability and process speed are always relevant criteria for their assessment. In the case of comparable product performance or a high degree of uncertainty or flexibility, short timescales play a prominent role. The strong emphasis on the process view in company management, especially in logistics, is its visible expression.

The broad methodical dissemination of process analysis or value stream analysis shows the deep understanding of the view of business activities in form of processes. Digitalization will further increase their relevance and the pressure on process efficiency.

The one-way street of shortening operational lead times is a predominant goal of management today. It should be emphasized that the structural fundamental decisions (see the management approach (i) to (ix) or Stalk/Hout) define the framework of process control and thus determine the possibilities of a possible shortening. The tools for reducing process time are derived from the above-mentioned principles.

A fundamental discussion about an increase of cycle time is very rare. For example, the idea of shortening the number of postal delivery days and thus increasing the buffer possibilities was only briefly discussed in public.

60

The digitalisation of data processes, which on the one hand leads to the automated overcoming of media breaks and interfaces and on the other hand to the automated transformation of data into information, allows process accelerations and adjustments in the organisation of material processes.

As a result, the paradigm of operative time management says: Faster or shorter is better, respectively.

5. Pragmatics: The paradigm in strategic management
a. Quote: If we don't act quickly, the game is lost.

This quotation has been reprinted as the title of an interview by Evgeny Morozov (see Morozov, 2017). He argues that the central question is not efficiency, but whether it is even possible to "stay in the game for the long term", i.e. to maintain or take on a sustainable competitive position. The interview discussed is about catching up from a weak starting position. In both cases "time" becomes the decisive yardstick; it determines victory or defeat and thus becomes a strategic success factor.

In the following, the question of how these strategic temporal effects were integrated into strategy concepts will first be examined.

b. Time in the leading strategy concepts

The discussion of "time" as a strategic success factor is first discussed on the basis of the outstanding strategy researchers who were honoured in the Thinkers50 Ranking (see the winners of the "Oscars of Management Thinking" at www.thinkers50.com) or who played an outstanding role in this ranking.

The early works of the strategy researchers Porter and Prahalad/Hamel did not consider "time" as a starting point, but their concepts can also be used for time-based competition: Porter (1980), with his '5 Forces', his 'Strategy Matrix' and his 'Value Chain', has established the basis for treating time as a relevant competitive dimension ('market-based view'). His strategy statements are based on the uniqueness of the operational services to the customer, and his concepts can basically be applied with the factor "time" as customer value. Thus, the quality in Porter's concepts can also be provided by temporal services. Kim and Mauborgne (2005) have conceptually expanded Porter's work with their "buyer utility map" by using the fundamental utility effects and the product life cycle as concept dimensions for a differentiation of services. Thus, the productivity of customers, their convenience or their simplicity in the phases of procurement or use are visibly improved by temporal services.

Prahalad and Hamel (1990), on the other hand, address the uniqueness of internal resources ("resource-based view"). They place the uniqueness of production and the creation of high barriers to market entry by making it difficult for competitors to copy services at the centre of their concept. In this respect, a unique temporal control of the production/distribution of services in the sense of Prahalad/Hamel can be interpreted as a strategic advantage.

For strategy researcher Collins, however, time does not play a prominent role because he extracts three explanations from his empirical studies for companies that are very successful in the long term: 'disciplined people', 'disciplined thinking' and 'disciplined action' (Collins, 2005, p. 20). For him, time is an implicit (and unnamed) conceptual component only through disciplined interaction.

Drucker, who also formulates successful strategies on the path to management quality by discussing five cornerstones of effective management, becomes clearer at this point: two of them have a direct reference to "time": (i) the correct use of time and (ii) the most important thing to do first (Drucker, 2014, p. 35ff. and p. 107ff.). The quality of management can be improved by the correct use of the resource "time" by giving strategic questions for managers the sufficient time frame necessary to develop and implement strategies. He argues for a diagnosis of time use, the consolidation of available time for these strategic issues, and the identification/ prevention of time thieves/wasting. The second component for Drucker, the prioritization, evaluates the possible strategic measures over time and requires on the one hand an evaluation of individual strategic measures and on the other hand a prioritizing approach from the executives. This means that Drucker does not contribute to strategic concepts in terms of content, but instead he provides an explicit and necessary prerequisite for effective time management. In this sense, his suggestions go beyond those of Collins.

Clayton Christensen et al. (2011) show how the classic strategic success factors with regard to customer, profit and growth orientation can have a critical impact on the existence of disruptive innovations. And this despite the fact that management has done everything right according to the 'rules of management'. But for them time only plays an accompanying role.

In summary it can be said that "time" in the concepts of the exposed strategy researchers is on the one hand fundamentally processable and treatable, but on the other hand does not represent the central core of their concepts. The question arises as to whether time is of the utmost importance or whether the concepts do not take this up to the necessary extent. In this respect, concepts will be examined that meet the demands of time as an explicit component of the concept and have not (as of today) yet been included in the Thinkers50 ranking.

A structured overview of strategies is provided by Scheuss, who has compiled 220 concepts in his book. It is remarkable that a chapter on "time strategies" is missing. There are only two sections where references become clear at the moment: (i) The outpacing strategy (Scheuss, 2008, p. 135ff.) describes the combination of price and cost leadership with market differentiation as a hybrid strategy. The market lead is always to be secured by changing the strategy content over time. (ii) Dynamic Capabilities (Scheuss, 2008, p. 147ff.) refer to the quickly adaptable 'intellectual' cognitive, analytical and implementation capabilities of management in order to cope with the rapid changes in the market.

Second, individual concept proposals for content time strategies have also been published. As examples may be mentioned Stalk (1989), Stalk / Hout (1990), Tunc/ Gupta (1993), Kumar/ Motwani (1995), Tersine/ Hummingbird (1995), Hum/ Sim (1996) or Ott (2011).

On the basis of the analysis of the term 'time management' in literature databases, however, it becomes clear that analytical work dominates "time" and that strategy concepts are still the minority. In summary, it can be said that strategy concepts with "time" as their core hardly exist or have been developed.

For this reason, the following section will thirdly demonstrate the strategic way in which "time" is or can be used in companies today. For this reason, no holistic strategy concepts are highlighted, but special time-based strategies. This describes short delivery times, high reliability, robustness/resilience and strategy of being on time.

c. Time strategies in the sense of short delivery times

In this section, time is considered in its effect as a Unique Selling Proposition (USP). This characteristic ensures that companies have a well-founded role from the point of view of purchasing companies or end customers, thus ensuring long-term sales. The uniqueness as a strategic core has been handled by Porter (1980) and Kim/ Mauborgne (2005). For the time strategy, the service provision of a company is then essentially geared to "time".

Time expresses the company's position in the market, i.e. vis-à-vis customers and competitors. Peters (1993, p. 105), for example, comes to the conclusion that in the future the basic strategic options can only be "fast or dead", whereby the second option excludes itself and thus only a "high speed strategy" remains as the sole lasting success formula for companies. This performance can occur in several ways.

Thus, time can be a product component that is relatively highly valued by customers. For example, the customer-related delivery time can be convincing as a decisive purchasing feature if the physical product performance on the market is comparable. The developments in distance selling with the promise of a delivery after 96 or 48 hours and now even a same-day delivery (SDD) show the strategic expectations directed at time competition.

A patent from Amazon (2013) can be cited as an initially futuristic example in distance selling: United States Patent No. US 8,615,473 B2 dated 24 December 2013 with the title 'Method and System for Anticipatory Package Shipping'. In fact, the packaging and delivery process begins before the individual customer places an order, and the short delivery time continues as USP. So it says in the summary of the patent specification:

„Various embodiments of a method and system for anticipatory package shipping are disclosed. According to one embodiment, a method may include packaging one or more items as a package for eventual shipment to a delivery address, selecting a destination geographical area to which to ship the package, and shipping the package to the destination geographical area without completely specifying the delivery address at time of shipment. The method may further include completely specifying the delivery address for the package while the package is in transit." (Amazon, 2013, p. 14).

Delivery time begins even before the customer places the order and thus provides the retailer with a competitive advantage in terms of time through the intelligent forecasting of customer orders. In extreme cases, the time of ordering and delivery overlap. For the strategist, however, the question arises as to where the shortening of this time competition comes to a 'natural' end.

The ILIPT project (Intelligent Logistics for Innovative Product Technologies) can be named as an industrial example of a strategic delivery time position. This project, funded by the European Union, aimed to develop a technical and logistical concept for the order-related production of an automobile within a defined delivery time of 5 days. The starting point is the current production system in the automotive industry, in which the final assembly as the core is always utilized to the same level and the existing Build-to-Order (BtO) orders are filled by Build-to-Stock (BtS) orders to ensure the continuous utilization of the final assembly. This results in either high speculative final inventories or longer waiting times for end customers. The aim of the ILIPT project was to guarantee a delivery time of 5 days and not to generate any stock after final assembly. In principle, the concept is based on the pillars:

- A new product structure with modular components and standardized non-brand specific parts

- A new network of suppliers with coordinated distribution of order penetration points (OPP) and flexible utilization of capacities according to incoming customer orders.

- A new logistics structure with direct connection of all participants to a "Virtual Order Bank" with integrated order, automatically generated production and respective capacity management.

- A new design of logistics processes with service rules (replenishment control) according to OPP and order situation.

This enables the car manufacturer to meet individual customer requirements while at the same time ensuring the delivery time target of '5 days' without an unnecessary build up of speculative stocks.

This ILIPT project can create individuality from the combination of standardized parts by designing value creation networks. The company DELL realized this years ago for the product 'Computer': no stationary trading points, but orders only via online orders, order acceptance, resolution of the individual order according to the parts list and forwarding of suborders to nearby suppliers of components with stock-keeping warehouses and short replenishment times, technically simple assembly processes and shipping options via a European parcel service with distance-independent prices. In this way, 'individualised' computers with a manageable number of parts could be delivered to customers throughout Europe within a week.

In both cases, ILIPT and DELL have succeeded in combining individual customer wishes with short delivery times. In both cases, these wishes form the starting point, which is carried out by a pull process chain without significantly higher unit costs. The contradiction has thus been resolved by logistics networks and transformed into a marketable and customer-oriented time strategy.

d. Time strategies in the sense of reliability and punctuality

The reliability of a process duration provides a planning benefit for the customer(s), because on the one hand they do not need to check the quality of the (service) performance of the provider (e.g. permanent checking of the quality) and on the other hand they do not need to take any arrangements to avoid the disadvantages of unpunctuality (e.g. additional warehouses or quick transports). This ability to rely is an elementary component of the just-in-time supply chain, because without punctual delivery they would experience permanent interruptions in production or rescheduling.

The discussion on punctuality reached Deutsche Bahn in the spring of 2019, whose statistics show that local trains left on time at 94.5 percent (1st half of 2018) and long-distance trains at 77.5 percent (1st half of 2018). Punctuality is defined as a delayed arrival or departure time of a train of more than 5 minutes and 59 seconds. In contrast, the Japanese high-speed train Shinkansen achieved an unpunctuality of less than 30 seconds (!) per train in 2013. In Japan, the punctuality is defined as 1-minute delay. Delays of more than 30 minutes are also said to have become the subject of the main TV news in view of the rarity. In Switzerland, on the other hand, a customer-oriented definition of unpunctuality applies, which means that only delays of a customer with more than 3 minutes at the destination are reported as unpunctuality. Follow-up punctuality is given as 97 percent (for all information see Zugreiseblog, 2019.

Punctuality ensures better planning and less effort (and annoyance) for all customers of delivery services for their own activities.

This phenomenon of cascading effects is visualized in Figure 5.1. It is assumed that there are three levels of trains: long-distance, regional and local, or comparable to a supply chain from Tier 2 to OEM. The travelling time corresponds to the transport duration and the waiting time to the logistical handling time. The trains (deliveries) depart on a clocked basis at all levels. If punctual, the turnaround times (waiting times) are sufficient to reach the next train (connection).

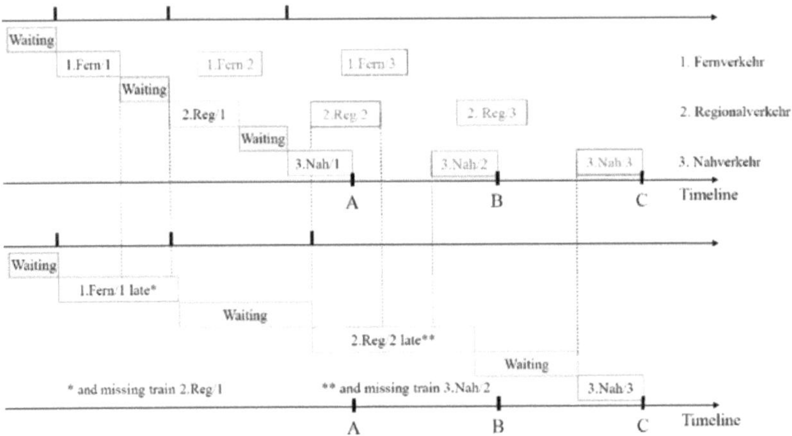

Figure 5.1: Punctuality cascade

In case of punctual arrival of the trains in the upper part of the figure, it can be seen that the connections are always reached and arrival at the destination at time A is possible. If connections are missed due to a delayed arrival of the train "1.Fern/1", only the following train "2.Reg/2" is reached. Also its delay leads only to reaching the next but one clock pulse "3.Nah/3" in the third level, i.e. only the late time C at the destination can be realized. In this example, the total delay of the delivery (customer) covers the time span from A to C. The respective delay cannot be compensated by the waiting times (buffer times) between the cycles.

This example of the unreliability of a three-stage transport serves to illustrate that the reliability performance of a stage can work through to the end. A positive counterexample is provided by Zara's value chain, which has a reliable textile pearl necklace based on its design, a cross-functional decision-making body, suppliers in close proximity and raw material warehouses (for example, Dutta, 2002). Here, not only the timing is remarkable, but also the control of new orders in view of the demand situation in the markets. Risk-reducing order sizes, speed in design implementation and delivery, and fashionable timeliness create a unique bundle of services.

All in all, these examples show the great entrepreneurial importance of looking at the entire supply chain for overall reliability over time. As a result, logistical analysis in the context of supply chain management have a greater entrepreneurial significance with regard to the impact on customer benefit.

e. Time strategies in the sense of robustness and resilience

Thirdly, time can also play a strategic role in the market within the framework of robustness or resilience. Robustness is resistance to external effects, so that these companies do not need to change their strategies for a longer period in the case of environmental changes. The resilience is the re-establishment ability for the case of a crisis/disruption/supply chain interruption. Such companies are characterized by the fact that they can register negative disruptions faster, analyse them faster, make faster decisions and implement them faster. These two attributes of time (robustness and resilience) require organisational, production-technical and logistical preparations.

Strategically designed networks, which prove to be robust, generate a double benefit for the company and consequently a strategic benefit for the customer: It is not necessary to adapt one's own strategy (business model or network) to changes in the environment and no measures have to be taken to re-establish the disruption. The customer can rely on reliable performance. The customer impact has already been explained in the previous section. Now the execution of the service carried out at the provider is to follow.

Figure 5.2 shows the individual time steps. In the period from t_0 to t_1, the provision of services is exposed to exogenous effects that do not cause a disruption or inefficiency of the business model. The last influence at time t_1 now interrupts the production process, for example by a fire in a company building or leads to a strategic crisis due to changed factor prices (inefficiency in the market). The performance level of the offering company drops significantly. The decision-makers are informed of this service interruption with a delay (t_1 to t_2), who then begin to prepare a decision, discuss alternatives and select a procedure (t_2 to t_3), which is then prepared for implementation (t_3 to t_4). In the period t_4 to t_5 the measures begin to take effect and only at time t_6 the original performance level is achieved again. The duration from t_0 to t_1 can be described as a robust duration and the duration from t_1 to t_6 as a resilient duration (for robustness and resilience see, e.g., Melnyk et al., 2015).

In a strategic discussion, fundamental preparations must be made in order to maintain a robust state for certain influences on the one hand and to restore the original performance level on the other. Time strategies for both situations have to be made.

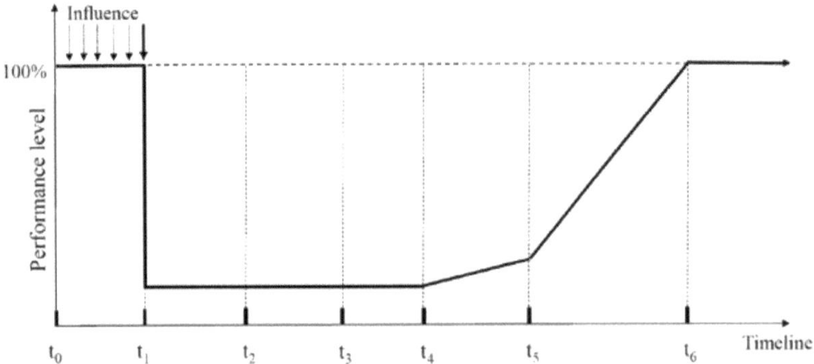

Figure 5.2: Robustness and resilience

The strategic measures for robustness require a risk management that identifies and evaluates the main types of risk for a company and avoids, reduces or shifts their negative impact through measures to be taken in advance. The time strategy is expressed by the duration of the neutralising effect with regard to exogenous influences. In view of changing economic conditions, it must always be questioned anew. A renouncement of such a strategy in the sense of robustness is of course possible; this is connected with the influence of (almost) all exogenous influences on one's own willingness to perform.

The robustness over time can also be an expression of the trust that customers have built up over a longer period of time. This characteristic cannot be built up at short notice (not even by competitors) and is one of the long-term competitive advantages of a company. It is expressed in an excellent reputation or strong brand strength and secures the longer-term market position (and sales). On the other hand, it is confirmed every day by outstanding performance.

The strategic measures with regard to resilience vary according to the individual phases. What they have in common is that they create a parallel organisation in order to provide services under difficult conditions in a complete management process:

-   t_1 to t_2: The information supply of the new framework conditions must be created in order to be quickly (!) informed about the new situation and its consequences. Thus, the precautions of possible infrastructures for data collection and data transmission have to be taken.

70

- t_2 to t_3: The possible reaction must be prepared and evaluated in a comparative manner; also, with regard to its temporal lead time and its temporal effect. This is to prepare the meeting of the decision makers and their information supply.

- t_3 to t_4: The preparation of the implementation of the decided measures should take place quickly (!). This means that coordination processes and decision-making processes must be practiced under special conditions.

- t_4 to t_5: The duration of the effect of the initiated measures should be short (!). This means that implementation steps must be trained.

- t_5 to t_6: The time of the restored ability to perform should be reached quickly (!). This means that measurement systems must be installed.

Even if these measures appear operational, they are still of a strategic nature, since they contain the basic willingness to perform in the sense of an infrastructure, within which the processes can run quickly in the event of a crisis.

f. Time strategies in the sense of timeliness

The fourth time-based strategy considers the timeliness of the services offered. It discusses so-called "window of opportunity", in which service plays a much greater role than outside this timeframe. This window of opportunity describes a particular period with high(er) strategic relevance or customer value.

Examples are the seasons when certain products are in high demand and no demand is generated, e.g. the Easter period with the Easter markets or the pre-Christmas period with the Christmas markets. These new time windows can also open up in fashion with seasonal changes. Such time windows also represent fairs at which new products are shown and at which these new products are also expected by visitors and retailers. The Mobile World Congress in Barcelona 2019, for example, is a time window during which smartphones with foldable displays and support for the 5G standard can be given a boost of attention and "pushed" into the minds of consumers.

This strategic timeliness requires companies to set up a value chain that is disciplined over time (better: development chain for innovations) in order not to miss this window. Otherwise, there is a risk of at least one year behind the competition, which will also be reflected in lower sales and earnings.

This ability to dominate the market requires the organisation of research and development services which should be carried out more quickly than competitors ('time-to-market'). These faster companies thus secure the role of pioneers and, as a rule, higher returns or better market exploitation (see the studies mentioned in Simon, 1989, p. 83ff.). The current development races for the best smartphone, the best laptop or the e-mobile with the highest distance prove the high relevance of this strategy.

The same decisions also apply to innovations as to production services: make-or-buy and, in case of a development based on the division of labour, the supplier structure (with whom?) and the form of strategic cooperation (preferred supplier or preferred customer) must be defined (see Darr, 2017b, p. 29ff.).

g. The evaluation of time-based strategies in the discussions of industry 4.0

The "time" factor impacts the development of industry 4.0 and the digitisation of value chains. Every time strategy is therefore influenced by this digital design. The starting point for this is formed by the respective market strategies, which are supported or shaped by the design of digitisation (prerequisites, e.g. machine readable data) and digitalisation (digital process organization) as "enablers". The respective data and information management is then to be designed strategy-specifically.

-   Delivery time strategies require a value creation network for coordinated serial process management. The digital possibilities enable a timely execution of the processes, their evaluation and their self-control in the face of supply chain disturbances (resilience).

-   Delivery reliability strategies connect to delivery time strategies by focusing on punctuality in the sense of a time guarantee. Here, too, the real-time status, its evaluation and its possibilities of self-control are the starting points for digitisation and digitalisation. The links between the activities in the process with regard to the quality of the overall service can be seen and controlled by digital elements.

-   Digitalisation provides robustness and resilience through improved time performance, makes the supply chain more robust, enables critical situations to

72

be identified and evaluated more quickly, and enables alternatives to be identified more quickly. A video of riskmethods (2017), in particular its second part, may serve as an illustrative example.

- Timeliness strategies should not be outlined so clearly in their support through digitisation: The evaluation of time windows that go beyond repetitive calendar opportunities requires intuitive and far-sighted assessments in order to be able to perform at the time of opening this possibility. The extent to which digital solutions will be able to recognize both the predictive quality for innovative framework conditions and the recognizability of the point in time is currently still open.

A fundamental discussion, here in the sense of a strategic discussion, on the potential of digitisation has been discussed by the Scientific Advisory Council of the Bundesvereinigung Logistik e.V. He has made the methodical approach and the assessment of the advantages of digitalisation in production and logistics based on eleven questions (see BVL, 2017, p. 5). Thus, the purely process-oriented view requires a questioning of the traditional goals of process control, the special role of normative and operational level, the coordination of participants in the value chain (particular interests) or the design of hierarchical coordination. Furthermore, the product, service and business model innovations, the qualifications and the role of risk and responsibility with regard to their need for change must then be put to the test. In this respect, the Scientific Advisory Board makes it clear that digitalisation should lead to a methodical and socio-technical change in corporate governance in order to exploit the expected benefits. The digitalisation is therefore more than a commissioning of additional hardware and software.

h. Interim conclusion

In all cases, it is evident that "time" as a service has become an integral part of companies' strategies and that the various strategy variants can be supported by digitalisation. For the time strategy in the sense of timeliness, this has yet to be developed. This development is supported by the comparison of product services in the global market and the organization as globally and efficiently connected supply chains.

It is also emphasized that "time" is not the only dominant strategic dimension but must be consolidated by empirical confirmations as a relevant effective dimension for profiling in the market for customers or supply chains.

It should be noted that unchecked continuation of once defined time strategies can also have significant disadvantages. This is done by comparing two quotations from the article by Stalk/Webber (1993): They make it clear from a strategic point of view that, against the background of the 1980s, time strategies already represented an effective way to gain distinct competitive advantages. On the other hand, they also make clear that time alone is not sufficient for differentiation in the market. They also point to a fundamental disadvantage of time strategies, which they call "Japan's dark side of time", i.e. permanent agility limits the possibility of stable returns. They therefore request that those responsible for strategy should master both types of strategy, i.e. position strategy and time strategy. They emphasize that the evaluation of time strategies has to be carried out especially with regard to customer effects and the internal possibilities of employees and technical performance.

The time strategy as the dominant corporate strategy is expressed by the following quote:

"Time offered a powerful new metric, a way of measuring performance in quality, variety, and productivity. And time was a powerful change agent. By looking through the lens of time, managers and workers could identify their organizations' main sequences, critical processes, and horizontal linkages, the elements that define the way a company does business and competes. [...] Time-based competition suggested an entirely new paradigm: the emergence of strategies of movement. [...] Time-based competition shifts the focus from the strength with which a company can create and defend a position to the speed and agility with which a company can move." (Stalk/Webber, 1993, p. 93ff.).

The time strategy as an integrated part of the corporate strategies is expressed by the following quotation in the same article by Stalk and Webber:

"Beyond the lesson of the three stages of time, Japan's plunge into the dark side serves as a reminder of something more basic: strategy is a constantly evolving exercise that requires relentless questioning on the part of managers. But the questioning should not be about strategy per se. To be meaningful, it should also be about customers and employees, about external wants and needs and internal capabilities and skills and the company's capacity to link the needs of its customers with the capabilities of its employees. As such, strategy can never be a constant; it must evolve. [...] Now the dark side of time suggests that a manager needs to be able to accomplish both types of strategy: strategies of movement to open up new

competitive space and strategies of position to cash in on the space once it's occupied." (Stalk/Webber, 1993, p. 93ff.).

Backhaus and Gruner (1997, p. 30ff.) also take a critical view of an exclusively accelerated time strategy by differentiating between the various negative effects: "straw fire effect", "crash effect", "pseudo-growth", "acceleration-resistant sales slide" and "conditions trap". Fülgraff (1997, p. 54ff.) emphasizes in his contribution the enormous natural disadvantages in the environment due to the fast pace of the industrial society and calls this therefore "surrender of the environment".

Despite the controversial assessment, developments in globalization and digitalisation have hardly changed since the 1990s despite changing general conditions. In this respect, today's contribution of strategic time management would have to be appropriately titled "bright side of time".

As a result, the discussion on the paradigm of strategic time management is: Time strategies create additional potential for differentiation and/or an additional cost reduction potential on the market. As a result, the business models are aligned under the maxim of acceleration.

6. Pragmatics: The paradigm currently in society

a. Quote: The time trap or hamster wheel looks like a career ladder from the inside

This frequently used quotation should clarify that a planned improvement of one's own situation (represented here as a career ladder with the ascension rungs) leads to no progress, if these necessary activities and processes are performed in a hamster wheel.

In order to better understand this quote, the statements are categorized according to the four levels of service: input (input as performance), throughput (process performance), output (outcome of processes as performance), and impact (positive effects and customer reviews), (Darr, 2017a, pp. 57f and Weber, 2012, p. 139). The fast race or the increase in speed therefore refers to the accelerated throughput. Climbing the steps of the career ladder leads to an increased output, but without achieving an additional effect, since the position in the hamster wheel is always the same. Every single one of them is pedaling ever faster, climbing further steps of the ladder and standing in the same place in the hamster wheel. The desired acceleration is de facto a time trap, because the individual does not develop further. This picture also corresponds to the interpretation of the concept of "furious standstill". (Virillo, 2015). Consequently, this time trap must be examined for the facts which do not make each individual leave the job.

The paradigms of acceleration of operational processes on the one hand and strategic competitive advantages through time strategies on the other are embedded in a larger context of social coexistence. According to the quote chosen, the same position in the fast-moving hamster wheel is always the metaphor for a lack of social development. For this purpose, the development of syntax, semantics and pragmatics are now classified on the basis of the acceleration categories of Hartmut Rosa, in order to then outline possible ways out or effective strategies in the sense of a genuine further development.

b. Acceleration means estrangement

The previous statements have shown that

- in the period of origin, the clock and the calendar have emerged from the rhythm of life with nature (nature-time),

- in the centuries that followed, adjustments were made to the solar calendars,

- the efforts of the revolutionary calendars did not last long,

- industrialization led to a standardization of the various regional times after a longer period in globally coordinated time zones and their measurements,

- time is a relevant dimension of operational processes and strategic market positions and will become increasingly important as a result of digitisation,

- the one-way street of a shortening of time as process time and customer value is paradigmatically pursued as the only way out and

- the economic developments of acceleration also have and will have social consequences.

The unreflected daily slogan is "you must always be faster" in order not to lose the connection in the competition or to get the advantage you hoped for. A circular acceleration, as described in the hamster wheel, is the result. Acceleration in operational and strategic management and in society is an observable fact.

Table 6.1 summarizes these developments using the three elements of semiotics and using the three generations in Rosa in a matrix of 9. The three generations are designated with 'yesterday', 'today' and 'tomorrow'. Thus, on the one hand, Rosa's view of society can be traced in the three columns of the table. And on the other hand, the development of syntax, semantics and pragmatics can be illustrated in the three lines of the table.

Thus, it becomes clear for the syntax that the development of the natural language of time over a standardization of the signs of the "clock time" (Levine, "Uhr-Zeit") now the complete coverage of daily processes was accomplished. At the level of semantics, the shift from the importance of natural rhythms over time to operational and strategic value has now taken place at all times. In the field of pragmatics, the shift from the importance of nature as a rational entity for the various cultures to the globally uniform economic paradigms of time management has taken place. The current development of hyperacelerated societies is currently in progress. Against this background, Hartmut Rosa developed his conception of acceleration as a characterization and demarcation of societies. The technically accelerated innovations continue unabated: The aforementioned quote to the Mobile World Congress "the discovery of speed" expresses this in the form of a headline.

	Yesterday	Today	Tomorrow
Syntax	- Watch: sun, sand, potato, etc. - Calendar: Julian, Gregorian, etc.	- Clock: International Atomic Time (TAI), Global Time Zones, UTC, ISO 8601 - Gregorian calendar until the year 9999	- Real-time screening and social evaluation: body cam and social points, also with regard to time
Semantics	- Rhythm of Nature - Nature Time - Event Time	- Clock Time - Time is a value for companies	- Timeliness and immediacy as value - Total feedback
Pragmatics	- P-time and M-time - Past-oriented, present-oriented and future-oriented culture - Long-term and short-term thinking	- Operational paradigms regarding process of value creation - Strategic paradigms regarding USP	- Furious standstill, timelessness of time, acceleration as basic maxim of life - Network economy

Table 6.1: Overview of the semiotic developments of "time"

The natural end of the shortening of the processes or the acceleration does not seem to have been achieved today: Here, for example, the patent by Amazon can be mentioned, where the picking and packaging is made in time before the actual final purchase decision of a single customer. The advantage of this patent, which includes a better customer-data-based forecast for the acceleration of delivery processes, is firstly valid a priori and has to assert itself only in the day-to-day planning world of individual customers. In the continuation of the logic of the hamster wheel this temporal preference could be made in the extreme for consumer decisions up to the birth hour. The absurdity of this statement (also given the self-determination of life) and thus the paradigm of speed are obvious.

It leads to an unlimited acceleration of our society. The increasing acceleration now leads in the extreme to signs to the dissolution of social values in "today". The speed of data knowledge and the speed of reaction require an all-encompassing data acquisition and an enormous speed in decisions and their implementation. In view of the current discussion on social media, the hunger for data does not need any further evidence. And the fast realisations lead to the 'temporary workers' who work on call and on an hourly basis. Here, too, a glance at the public media or newspapers is enough. This detachment from traditional values (see Rosa) has already been described by Postman under the title "Technopoly: The Surrender of Culture to

Technology". From an economic point of view, Rifkin (2014) explains the consequences in his book "Zero Marginal Cost Society".

For a 'tomorrow', all social processes could be documented in real-time (screening) in order to evaluate the conditions or behaviour of individuals, groups or companies with regard to defined overarching national goals and to control and influence this behavior. The introduction of social points and access to credit or the real estate market in China show what this could look like. The valuation is carried out according to the claim of immediacy or timeliness. The discussion on the availability of executives by e-mail at any time shows the relevance of this discussion. Rosa (2016a) discussed the pragmatic consequences extensively. The social discussion of excessively time-driven societies has become clear with Michael Ende (Momo und die Zeitsparkasse), in Charlie Chaplin's film (Modern Times and Screw-Fastening) or in the film 'In-Time' (paying for consumer goods with a lifetime). The increase in burnout cases may be cited as another indication of accelerated societies.

Both developments, growing digitisation and digitalisation of social life, increasingly create the technical prerequisites for an all-encompassing collection of data and persons, and thus also a way of processing these data with regard to socially standardized orientation and regulation. To what extent social media will influence the emotional connection and the emotional development of shared experiences will be the subject of future studies. The formation of common values and their social benefits are discussed in Section 6d.). According to the thesis, reference-based experiences for future social behaviour and their further development are not encouraged; but on the contrary even limited.

The increasing alienation as a result of technical and social acceleration can already be observed several times today. It raises the question of what the next steps can look like if a 'continue like this' is to be counteracted or the development is to be reversed.

c. Deceleration and idleness as ways out?

If this unrestrained acceleration promises no benefits in the long run, then a reversal, i.e. deceleration, is an obvious possible answer. This idea is not new and was published several times as guide for deceleration resp. meaningful way of life. A small selection is briefly presented:

Lothar Seiwert has produced a number of guidebooks on this subject. His 'Das 1x1 des Zeitmanagement' (Seiwert, 2014) is a standard book for personal time management. It helps to examine its use of time, to set its goals and priorities, to formulate strategies for own time and to develop its goal consciousness. Personal time is then 'managed' in the same way as professional time (see also Dobbins/Pettman, 1998 or Rao, 2014). Also, the association for the delay of time (www.zeitverein.com) has dedicated itself to the conscious time balance and/or deceleration. Seiwert's book 'Bären-Strategie' (Seiwert, 2005) formulates the claim "Bärig lebt es sich besser!" and discusses the interplay between calm and dynamics. His books "If you are in a hurry, go slowly. More time in an accelerated world" (Seiwert, 2018) and "Let go and you are the master of your time" (Seiwert, 2013) focus on personal time sovereignty, work-life balance and personal happiness. Thus, he started his concepts with the efficient use of time and continuously develops procedures to make life meaningful.

Holm Friebe's book 'Stein-Strategie' (Friebe, 2013) deals with the conscious decision to use time. He describes the implementation and achievement of one's own goals/interests by waiting and not acting. In an interview with him by Franz and Claudia Mattig, these statements are mentioned as follows:

"In our time, there is an aptitude to run after trend topics or jump on every idea that is run through the roost. It is not paused, reflected, and considered, whether the wise waiting in this situation could also be a probate strategy. [...] «When you move, you need to know where to go.» For me, that's the definition of a strategy. [...] «If you do not move, you need to know why.» [...] The true art is to discipline oneself and keep a decision open as long as possible within the time window. This approach may have to do with experience, with age, with maturity - but not necessarily. But it is the only strategic recommendation that I can derive from my preoccupation with this matter." (Friebe, 2015).

Deceleration as a reversal of acceleration can also be found in consumer products. As an example, slow food can be mentioned. 'Slow TV', which began in 2009 with the Norwegian channel NRK, also shows a train journey of several hours between Bergen and Oslo. There is no dramaturgy of a film script, but the possibility of experiencing the train journey in full length uncut. In the media library of Bayerischer Rundfunk, 'MORA' is a Slow TV format in which people are observed for 45 minutes while working. Examples include the jewellery designer, the ice sculptor, the cello maker and the watchmaker.

Futurologists and their institutes have also addressed this issue of 'deceleration': 'Zukunftsinstitut (The Future Institute)', for example, has included other categories on its homepage that deal with slowing down life's activities: (a) The Age of Slowness, (b) Slow Thinking, (c) Slow TV and (d) Slow Management. The list of links can be found in the bibliography under Zukunftsinstitut (2018a to 2018d).

Such offers and advisors can be found, which set consciously on the deceleration in the hectic world of the "timeless time" and thus in the core a more rational handling of the own use of time on the one hand and a product of the warmth, the community and thus a product of the demarcation to create the hectic and restlessness of life on the other hand. The discussion on the counterpart of acceleration is therefore taking place; in short: no trend without countertrend. These self-limiting behaviors are based on individual choices. To what extent these will lead to a majority reversal of social acceleration is currently unforeseeable.

d. Resonance or self-organization?

Here, the focus is on two proposals to stop or counteract the "slippery" development of acceleration. These are the proposals (i) by Hartmut Rosa, who also designed a prognosis model or a solution following his explanatory model (see 1e.), and (ii) by Peter Kruse with his concept of self-organization. First the prognosis model by Hartmut Rosa:

On the basis of his analyses and arguments for unchecked acceleration, he initially considers four scenarios to be justified (Rosa, 2016a, p. 486ff.).

Scenario 1: Social, political and legal institutions are formed at the higher acceleration level that has been reached, which are capable of comparable organizational and orientation performance as in the 'classical modern'. From Rosa's perspective, this scenario is not probable.

Scenario 2: The process of acceleration is controlled by a new kind of (sub)politics, without Rosa becoming more concrete here. An end to the acceleration trend is not foreseeable here.

Scenario 3: Forced deceleration through a politically initiated emergency brake ("revolution against progress"). From Rosa's point of view, this scenario is highly unlikely.

Scenario 4: An unchecked continuation takes place in an abyss that is reflected in the form of political, social and economic catastrophes.

For Rosa himself, the alternatives mentioned are "an extremely disturbing end". (Rosa, 2016a, p. 490). In this respect, he opens up the possibility of an additional proposal, which he later published under the title 'Resonance': "If acceleration is the problem, then resonance is perhaps the solution". (Rosa, 2016b, p. 13). His thoughts are based on the desire for recognition and a way out of alienation in order to avoid the pathologies of time. His approach is based on an intrinsic motivation of the individual to escape this acceleration trap and recalls two categories of persistence from his monograph 'Acceleration': on the one hand the deceleration islands, in which cultural islands are created against the time trend (anachronistically) (Rosa, 2016a, p. 143), and on the other hand deceleration as ideology, in which the individual withdraws from the pressure of acceleration as a drop-out (even if he accepts a high price) (Rosa, 2016a, p. 146ff.). The extent to which the resonance proposal represents a solution for the mass of participants in a society may also be discussed critically, since the causes of technical and social acceleration are not the core of his proposal. Also, the political acceleration criticized by Rosa (2016a, p. 391ff.) remains outside the resonance discussions.

Now to Peter Kruse's proposal for self-organization:

Kruse (2011) has written in his book "next practice. Successful Management of Instability" presents his ideas of successful management in complex and unstable environments[Fn6]. He proposes and argues that the principle of organisation is the so-called self-organization. Among its core elements are the entering into a phase of instability, the repeated analysis of the results of continuous decision sequences (through iteration) and thus the formation of network intelligence. The purpose and use for enterprises consist in the justified answer to external and internal basic conditions as new formulation of a lastingly successful strategy. At Kruse, this is referred to as a change of pattern. This presupposes cultural change and system competence on the part of management (Kruse, 2011, p. 153ff.).

It plays an important role for him that stable behaviour patterns are consolidated by culture and hierarchies, and that changes or strategy changes are always attacks on established behaviour.

Kruse examines the necessity of a change in strategy, despite all possible resistance, on the basis of five characteristics: changes in the market environment, changes in work organisation, changes in the corporate structure, changes in product strategy and changes in management. If such symptoms are present, a process of instability must be initiated. The management level's commitment to this instability and to temporary losses, as well as the emotional resonance of the vision, can initiate this process in the long term. The acceptance of this new strategy (vision) by those affected (here: employees) presupposes the credibility of the management. A convincing communication of the need for action is only one aspect of successful strategy work. Furthermore, by the developing feedbacks (iterations) the new strategies are permanently examined, reflected back to the management and they contribute thereby self-reinforcing to the acceptance, to the advancement and to the success of the enterprise.

Peter Kruse derives his concept from the findings of brain research. The interaction of evaluation (limbic), activation (reticular) and innovation (cortical) allows him to make intuitive and holistic model evaluations even in complex and unstable situations. Through the interaction of the three characters owner (rating), creator (innovation) and broker (activation) unique achievements are created that are problem-adequate and effective: The combination of Broker & Creator leads to a lively exchange of new ideas. The combination of Creator & Owner leads to evaluated ideas (innovations) and the combination of Owner & Broker leads to a lively exchange of evaluations, so that the individual actors can check whether these are helpful for them. Exactly these characteristics come to bear in self-organization: No linear hierarchical processes and no monocultures, but the free access (characteristic exchange) to new ideas (characteristic novelty), which are competently assessed for their effectiveness (characteristic assessment) and implemented (iteration).

An essential characteristic for Kruse is intuition, i.e. the spontaneous holistic perception and evaluation of complex situations. These intuitive patterns, which are crucial for the complex assessment[Fn7], are only formed in the brain through prolonged work in the border area (crises) and through operative responsibility. Intuitions are the guarantor of an effective evaluation of new strategies.

Kruse transmits these mechanisms of brain performance to companies. It is an inversely image of the three elements of the brain (owner, creator, broker) to precisely select these three characters for an organization, to initiate their interaction in the organization and to allow the free interaction between these three characters.

Kruse then proposes a pragmatic action plan for companies in unstable and complex situations:

- To promote the creation of a common culture within the company, e.g. through shared emotional experiences and situations.

- Increase networking between employees, i.e. "full access", e.g. through free self-responsible exchange of ideas and information.

- Clarification of the added value of a networked group contribution by each individual, e.g. through clear and widely accepted evaluation criteria.

- Allowing intuitive assessments, e.g. by accepting individual assessments rather than endless justifications through analysis. However, these judging persons should have experienced crises to have formed their ability to judge or intuitions.

- Formation of quality criteria for feedback, e.g. intuitively influenced responsible persons, determine the evaluations with regard to the effective future strategy.

- Management committees only evaluate the solutions, e.g. through open discussion on the basis of widely accepted evaluation criteria. Leadership does not contribute to their development of the solution portfolio.

The exciting aspect of Kruse's proposal is the interplay between the various hierarchical levels and the controls of leadership in complex and unstable environments. Executives have a distinctive and proven ability to judge, which they have acquired through 'crises' and thus gain acceptance among employees. The developed guidelines (rules) provide the framework within which the employees can and should decide freely. Feedback is intended to control.

This basic logic can now be transferred to the control of societies. The corporate leadership here corresponds to the political leadership. Rosa's criticism of them would not come into play at Kruse, since the intuitive ability to judge ensures the necessary degree of fast and accurate control. Thus, different standards and different selection procedures are used at Kruse for the management than at Rosa for the concept.

In addition, day-to-day processes do not run 'just like that' and slide into an accelerated future. The retarding elements are expressed through the above-mentioned framework and the evaluative feedback in the form of a contribution to the whole. Thus, the whole, i.e. society, is always held together by common value contributions. The emergence of these shared values is the central responsibility of political leadership. Their art then is not to frame too narrowly or too broadly so as not to stifle initiatives or let them run into the shoreless. Kruse's answer to this is the intuition and open discussions or feedback of the three characters who keep the process going. As in the brain, new, situation-appropriate, value-contributing and accepted ideas for further development emerge.

With his proposal, Kruse addresses precisely the core of the problem of accelerated societies. He takes up a categorization of Mintzberg for the situation-adapted coordination of stable/unstable/simple/complex situations and transfers the control logic of the limbic system to companies in a turbulent environment, which can also be transferred a priori to societies. It provides a proposal related to the core of the problem 'acceleration of and in societies' and an analysis and decision logic proven and developed by nature.

However, the question remains as to how the current established mechanisms of leadership and control, which matured in earlier stable times, can now develop independently in the direction of self-organization.

The factor "time" and its social consequences are undisputed. Starting from Rosa's model of society, who with his concept of acceleration has defined generations and thereby also describes the consequences of the ever faster rotating hamster wheel, the numerous personal guidebooks also show that stress, alienation, the lack of self-determined life, the lack of time or the lack of commitment have arrived as powerlessness or weakness with the individual. This is in view of the unrestrained attention of the topic also no flash in the pan.

Conceptually, Rosa has taken the starting point for the negative effect on those affected. Thus, the cause remains in the medium term and the encapsulation in an oasis will help to alleviate or avoid the condition for individuals. De facto, however, only sufferers leave the circulatory system without changing it. Kruse has started at the cause. It remains to be seen to what extent observable measures for change can be implemented here; however, at least on the intellectual level, the train of thought seems more problem-oriented and more effective in the long term.

e. Final Conclusion

The final conclusion is presented in Figure 6.1. It shows that the retrospective development of clocks and calendars was initially oriented towards cultural and natural conditions. Also, the culturally different social developments of the peoples and nations count to it.

Different historical developments of the peoples / countries
 	
From living in harmony with nature to the industrialized world
	• Calendar in alignment with the orbit of sun and moon
	• Change to the Gregorian calendar in the Western world
	• Development of watches according to technical possibilities (hours, minutes, seconds)
	• Culturally different developments:
	  - P-time/ M-time
	  - Nature Time / Event Time / Clock Time
	  - Past / present / future reference
	  - Long-term / short-term thinking

⇩

Changes	Consequences
• Industrialisation • Worldwide trade relations • Digitalisation	• Three functions of time must not only be fulfilled locally or regionally: Language (communication), evaluation yardstick/ assessment standard (orientation) and strategy (regulation) • Paradigm: Acceleration as a one-way street

⇩

Syntax	Semantics	Pragmatics

• Worldwide standardization of the language of time:
 	
ISO standard, UTC, time zones, calendar, atomic clock
	• Time as value
	  - Operative process efficiency
	  - Strategic unique market position (USP)
	  - Social acceleration: technical, social and pace of life
		• Shortening time as an operational and strategic paradigm
		• Self-organization as a possible path forward for accelerating societies

Figure 6.1: Summary of the phenomena, functions and paradigms of time management

Today, both the syntax and semantics of "time" are highly unified by coordinated globalization; partly also with regard to the social effects of all in the hamster wheel. According to the well-founded thesis, digitisation and digitalisation will have a further accelerating effect here. Effective forms of self-organization, such as the possible path to digital self-determination without the detrimental consequences described above, still need to be developed socially.

This "time-based" conclusion must not hide the fact that time is only "relating" with the social symbols of the clock and the calendar. Hence, on the basis of a global language of "time", collectively accepted comparisons, measures, and strategies can be made. Nevertheless, "time" is only a possible artifact with defined functions. In other words, it would be just as possible to communicate about facts without "time," unless interdependencies are discussed. If these requirements for a procedural coordination do not exist or if time does not play a strategic role, the clock and the calendar could basically be dispensed with. However, if there are temporal dependencies, then it requires a common medium, so that this coordination can be successful. This is shown e.g. by the historical examples themselves to ensure food production. Ultimately, the coordination and thus clocks and calendars are operationally, strategically and socially indispensable.

Thank you for your interest and your time.

7. Footnote

Fn1:.The newspaper Welt am Sonntag of 20 January 2019 provides a satirical account of the 'relationship setting' on page 10: The picture shows a conductor pointing to the new station clock without hands and emphasising this as a clear advantage, because annoyance about train delays will now decrease sharply. This is a satire and yet it also hits the core of the time discussion: without a pointer nothing can be related, and therefore no comparison of punctuality or delay can take place any more.

Fn2: The following information is provided for interested readers: Five societies are engaged in the study of time. They are named 'International Society for the Study of Time' (http://www.studyoftime.org) on the homepage of the Society: The Royal Society Philosophical Transactions of the Royal Society, Temporal Belongings Research Network, Deutsche Gesellschaft für Zeitpolitik, The Japanese Society for Time Studies and International Society for Philosophy of Time. The individual homepages then refer to further institutions and publications dealing with the study of time and its special phenomena. The homepage of the Deutsche Gesellschaft für Zeitpolitik is mentioned as an example with the mention of studies, books and magazines: http://www.zeitpolitik.de/international.html.

Fn3: "Tradition is basically nothing other than the recognition of the authority of symbols and the relevance of the narratives from which they have emerged." (Postman, 1992, p. 184).

Fn4: This wording was also chosen by the singer Unheilig as the title of his music album: "Everything has its time. Best of Unheilig 1999 to 2014".

Fn5: Information products express the value of data. There are four different status levels: **(1)** Information product 1 describes the provision of data. The existence of data is described as "machine-readable documented or *documented*". The facts of the process (or process step) are documented with regard to relevant features in machine-readable form and can be evaluated mechanically and promptly with regard to these features. The performance of information product 1 lies in its ability to provide historical data in a timely and individually situational manner, for example for purposes of context-specific evaluation, decision preparation or decision making. **(2)** Information product 2 describes the evaluation of the quality of data. This is referred to as "*checked*". This process (or process step) is then evaluated with regard to defined characteristics, i.e. checked. The performance of information product 2 lies in the (traceable) assurance of the quality of the upstream and current informational, material or space-time process performance. Information product 2 is generally based on information product 1 (acquisition and storage).

(**3**) Information product 3 describes the process result and is referred to as "*decided*". The process (or the sequence of the process steps) has been defined, i.e. decided. The performance of a decision here lies in the conscious release to continue the steps in the order cycle or the spatio-temporal added value, the improved control of the processes and/or the assurance of conformity. Information product 3 is usually based on information product 2 (analysis and testing). (**4**) The information product 4 describes the benefit for the subsequent user in the process and is described as "*effective*". The process (and/or the sequence of the process steps) has effective effects for the purchase or the supplier, e.g. in the form of guaranteed date keeping or stable incoming orders. Also, the possibility of the traceability of analyses and decisions is an example of an information product 4. The achievement lies here in the possibility of the time near subsequent examination and analysis (i) of the documentation of relevant characteristics of the creation of value processes, (ii) of the examinations made and (iii) of the decisions made for the continuation and control of the processes and/or the creation of value. The scope of the documented data of the processes, the states, the analyses and decisions define (limit) the possibilities of traceability.

The state of a more or less digitized process chain in the order cycle then ranges from state 0, state 1 (information product 1 is fulfilled) to state 4 (information product 4 is fulfilled).

The border between analogue and digital world is called "**Digital Penetration Point**" (red line). This can refer to the utility level "process", "result" or "customer" (for detailed discussion see Darr, 2017a and 2019) and is shown in Figure 7.1. In the digital world, interfaces and media disruptions are overcome by software applications (blue arrows); in the analogue world, employees perform this task (black arrows).

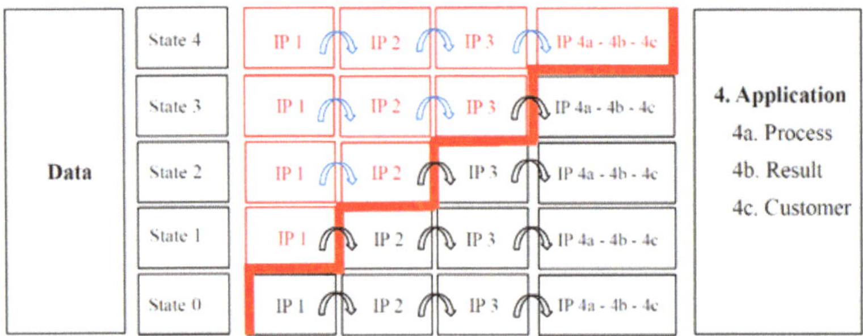

Figure 7.1: Digital penetration points

Fn6: The basis of his concept is the assignment of management concepts for stable/unstable and simple/complex environments (Kruse, 2011, p. 41). Ashby's law is used as justification. It states that the variety of the control system must be at least as great as the variety of the malfunctions that occur in order for it to carry out control effectively. In complex and unstable environments, a powerful control system is needed, which can only be achieved by self-organization. Furthermore, Kruse sees alternative management measures for engaging in the new strategy.

Peter Kruse derives his concept from the findings of brain research, because he is convinced of the superior performance of the brain through permanent powerful solutions. In this way, he transfers the functionality of the brain to the corporate management functions: Change in the brain (pattern change, strategy) does not come about through instructions, but through diversity and processes that are triggered by the inner tension of this diversity. These systems with internal tensions generate precisely these changes (unstable phases).

This diversity is represented in the brain by three appearing characters:
-    - The **Creator** generates emotions and new ideas.
-    - The **Owner** generates detailed analyses and evaluates
-    - The **Broker** creates links between actors.

Fn 7: In Kruse (2011), the characteristics of intuition include the unconscious capture, evaluation and decision of a complex context without being able to explicitly express the individual underlying relationships. Colloquially, this is known as "gut feeling". Measuring this collective intuition, however, is a methodological challenge for Kruse.

8. Bibliography

Amazon (2013): United States Patent No. US 8,615,473 B2 vom 24.12.2013 mit dem Titel ‚Method and System for Anticipatory Package Shipping, https://patentimages.storage.googleapis.com/8a/67/ff/299703230243b5/US861 5473.pdf, (Retrieved: 01.03.2019)

Armistead, Colin (1996): Principles of Business Process Management, in: Managing Service Quality: An International Journal, Vol. 6, Iss. 6, 1996, pp. 48-52.

Arte (2018): https://www.arte.tv/de/videos/068398-000-A/zeit-ist-geld/ (Mediathek: Ansicht am 4.11.2018)

Backhaus, K.; Bonus, H. (Hrsg.) (1997): Die Beschleunigungsfalle oder der Triumph der Schildkröte, 2. Aufl., Stuttgart 1997

Backhaus, K.; Gruner, K. (1997): Epidemie des Zeitwettbewerbs, in: Backhaus, K.; Bonus, H. (Hrsg.): Die Beschleunigungsfalle oder der Triumph der Schildkröte, 2. Aufl., Stuttgart 1997, S. 19-46

Becker, Torsten (2008): Prozesse in Produktion und Supply Chain optimieren, 2. Auflage, Berlin/Heidelberg 2008

Bögel, Stephan; Stieglitz, Stefan; Meske, Christian (2014): A role model-based approach for modelling collaborative processes, in: Business Process Management Journal, Vol. 20 Iss. 4, 2014, pp. 598-614

Burlton, Roger T. (2001) Business Process Management: Profiting from Process, Sams Publishing 2001, especially chapter 3, Principles of Process Management, pp. 65-97

BVL (2017): Logistik als Wissenschaft - zentrale Forderungen in Zeiten der vierten industriellen Revolution. Positionspapier des Wissenschaftlichen Beirats der Bundesvereinigung Logistik e. V. (BVL), Bremen 2017

Christensen, Clayton; Matzler, Kurt; von den Eichen, Stephan F. (2011): The Innovators Dilemma. Warum etablierte Unternehmen den Wettbewerb um bahnbrechende Innovationen verlieren, München 2011

Collins, Jim (2005): Der Weg zu den Besten. Die sieben Management-Prinzipien für den dauerhaften Unternehmenserfolg, 5. Aufl., Stuttgart/München 2005 (Englisch: Good to Great, Random House, London 2001)

Darr, W. (1992): Integrierte Marketing-Logistik, Wiesbaden 1992

Darr, W. (2017a): Digitale Transformation zum Einkauf 4.0. Nutzenbasierte Konzeptionen zum Smart Procurement, Hamburg 2017

Darr, W. (2017b): Spezialfragen des Einkaufsmanagements, Hamburg 2017

Darr, W. (2019): Procurement Excellence: Zum Leistungsprofil und zum Grad der Digitalisierung des Einkaufs. Forschungsbericht zu den Ergebnissen der Befragungen regionaler mittelständischer Unternehmen, Hamburg 2019

Delfmann, Werner (2010): Entschleunigung, Entkopplung, Konsolidierung – Ansatzpunkte für einen Perspektivenwechsel in der Logistik, in: R. Schönberger/R. Lebert (Hrsg.), Dimensionen der Logistik. Funktionen, Institutionen und Handlungsebenen, Wiesbaden 2010, S. 586-600

Dobbins, R.; Pettman, B. (1998): Creating more time, in: Equal Opportunities International, Vol. 17, Iss. 2, 1998, pp. 18-27

Drucker, Peter F. (2014): The Effective Executive - Effektivität und Handlungsfähigkeit in der Führungsrolle gewinnen, München 2014

Elias, Norbert (2017): Über die Zeit, Suhrkamp Taschenbuch Wissenschaft, 12. Aufl., 2017

Ende, Michael (2005): Momo oder Die seltsame Geschichte von den Zeit-Dieben und von dem Kind, das den Menschen die gestohlene Zeit zurückbrachte, Stuttgart/Wien 2005

Forrester, J. (1961): Industrial Dynamics, Cambridge, MIT Press 1961

Franklin, Benjamin (1748): Ratschläge für junge Kaufleute, im Original: Advice to a young tradesman, written by an old One, Works ed. Sparks, Vol. II, S. 87 (oder: www.franklinpapers.org, Retrieved: 25.02.1019)

Friebe, Holm (2013): Die Stein-Strategie. Von der Kunst, nicht zu handeln, München 2013

Friebe, Holm (2015): Interview von Franz und Claudia Mattig mit Holm Friebe, https://mattig.swiss/die-stein-strategie/ (Retrieved: 21.04.2019)

Fülgraff, Georges (1997): Entschleunigung, in: Backhaus, K.; Bonus, H. (Hrsg.): Die Beschleunigungsfalle oder der Triumph der Schildkröte, 2. Aufl., Stuttgart 1997, S. 47-66

Goldratt, E.M.; Cox, J. (2013): Das Ziel. Ein Roman über Prozessoptimierung, Frankfurt/ Main 2013

Hagen, N.; Nyhuis, P.; Frühwald, Chr.; Felder, M. (2006), (Hrsg.): Prozessmanagement in der Wertschöpfungskette, Bern/Stuttgart/Wien 2006

Hall, Edward T. (1983): The Dance of Life. The Other Dimension of Time, Anchor Books, A Division of Random House, Inc., New York 1983

Harrison, Alan; von Hoek, Remko (2008): Logistics Management and Strategy. Competing through the supply chain, 3rd edition, Prentice Hall, Harlow 2008

Hawking, Stephen (2018): Eine kurze Geschichte der Zeit, 19. Aufl., Reinbek bei Hamburg 2018

Horx, M. (2003): Future Fitness. Wie Sie Ihre Zukunftskompetenz erhöhen. Ein Handbuch für Entscheider, Frankfurt am Main 2003

Hum, Sin-Hoon; Sim, Hoon-Hong (1996): Time-based competition, in: International Journal of Operations & Production Management, Vol. 16, Iss. 1, 1996, S. 75-90

Kartoffel (2018): http://www.kartoffelhaus-hirschhorn.de/kartoffelhaus/kartoffelgeschichte/, (Retrieved: 29.01.2019)

Kim, W. C.; Mauborgne, R. (2005): Der blaue Ozean als Strategie. Wie man neue Märkte schafft, München 2005

Kim, Henry M.; Ramkaran, Rajani (2004): Best practices in e-business process management: Extending a re-engineering framework, in: Business Process Management Journal, Vol. 10, Iss. 1, 2004, pp. 27-43

Klepzig, Heinz-J.; Schmidt, Klaus-J. (1997): Prozessmanagement mit System, Unternehmensabläufe konsequent optimieren, Wiesbaden 1997, speziell die Seiten 87-109

Kluckhohn, F.; Strodtbeck, F.L. (1960): Variations in Value Orientations, Greenwood Press, Westport 1960

Kruse, Peter (2011): next practice. Erfolgreiches Management von Instabilität. Veränderung durch Vernetzung, 6. Aufl., Offenbach 2011

Kumar, A.; Motwani, J. (1995): A methodology for assessing time-based competitive advantage of manufacturing firms, in: International Journal of Operations & Production Management, Vol. 15, Iss. 2, 1995, pp. 36-53

Kotzab, Herbert; Otto, Andreas (2004): General process-oriented management principles to manage supply chains: theoretical identification and discussion, in: Business Process Management Journal, Vol. 10 Iss. 3, 2004, pp. 336-349

Lay, Thomas et al. (2012): Kriterien zur Leistungsbeurteilung von Prozessen: Ein State-of-the-Art, in: Multikonferenz Wirtschaftsinformatik 2012, Tagungsband der MKWI 2012, Hrsg.: Dirk Christian Mattfeld; Susanne Robra-Bissantz, Braunschweig 2012

Levine, Robert (1997): A Geography of Time. On Tempo, Culture, and the Pace of Life, Basic Books, A Member of the Perseus Books Group, New York 1997

Levine, Robert (2016): Eine Landkarte der Zeit. Wie Kulturen mit Zeit umgehen, 20. Aufl., München/Berlin 2016

Liebetruth, Thomas (2016) Prozessmanagement in Einkauf und Logistik, Instrumente und Methoden für das Supply Chain Prozessmanagement, Wiesbaden 2016, hier: S. 95-103

Luckner, Andreas (2012): Zeit und Existenz aus Sicht der Philosophie, https://www.uni-hannover.de/fileadmin/luh/content/alumni/unimagazin/2012_zeit/netz10_luckner.pdf, (Retrieved: 19.02.2019)

Morozov, Evgeny (2017): Interview von Dieter Schnaas in der WirtschaftsWoche vom 31.10.2017, https://www.wiwo.de/technologie/digitale-welt/techkapitalismus-handeln-wir-nicht-schnell-ist-das-spiel-verloren/20504708-all.html, (Retrieved: 01.03.2019)

Morris, Charles W. (1938): Foundations of the Theory of Signs, in: International Encyclopedia of Unified Science, Vol. 1, Number 2, 1938, pp. 1-59

Nadolny, Sten (2018): Die Entdeckung der Langsamkeit, 55. Aufl., München 2018

Nordsieck, F. (1934): Grundlagen der Organisationslehre, Stuttgart 1934

Nyhuis, P.; Grabe, D.; Nickel, R. (2006): Bewertung von Fertigungs- und Lagerprozessen mit Logistischen Kennzahlen, in: Prozessmanagement in der Wertschöpfungskette, Hrsg.: N. Hagen, P. Nyhuis, Chr. Frühwald, M. Felder, Bern/Stuttgart/Wien 2006, S. 79-116

Obermaier, Robert (2016): Industrie 4.0 als unternehmerische Gestaltungsaufgabe: Strategische und operative Handlungsfelder für Industriebetriebe, in: Industrie 4.0 als unternehmerische Gestaltungsaufgabe. Betriebswirtschaftliche, technische und rechtliche Herausforderungen, Hrsg.: R. Obermaier, Wiesbaden 2016, S. 3-34

Ohno, Taiichi (2013): Das Toyota Produktionssystem, 3. Aufl., Frankfurt/Main 2013

Ott, A.; (2011): Time-value economics: competing for customer time and attention, in: Strategy & Leadership, Vol. 39, Iss. 1, 2011, pp. 24-31

Peters, Tom (1993): Jenseits der Hierarchien, Düsseldorf 1993

Picot, A.; Reichwald, R.; Wigand, R. (2003): Die grenzenlose Unternehmung, 5. Aufl., Wiesbaden 2003

Porter, M. (1980): Competitive Strategy: Techniques for Analyzing Industries and Competitors, New York: Free Press 1980

Postman, Neil (1992): Das Technopol. Die Macht der Technologien und die Entmündigung der Gesellschaft, Frankfurt am Main 1992

Prahalad, C. K.; Hamel, Gary (1990): The Core Competence of the Corporation, in: Harvard Business Review, Vol. 68, Issue 3, 1990, pp. 79-91

Rao, M.S. (2014): Timeless tools to manage your time, in: Industrial and Commercial Training, Vol. 46, Iss. 5, 2014, pp. 278-282

Rifkin, J. (2014): Die Null Grenzkosten Gesellschaft. Das Internet der Dinge, kollaboratives Gemeingut und der Rückzug des Kapitalismus, Frankfurt am Main 2014

riskmethods (2017): The Future of Supply Chain Risk Management, https://www.youtube.com/watch?v=9nDZJyx3fq0, (Retrieved: 06.03.2019)

Roland Berger (2015): Die digitale Transformation der Industrie, www.bdi.eu/media/user_upload/Digitale_Transformation.pdf, Studie im Auftrag des Bundesverbandes der Deutschen Industrie e. V., (Retrieved: 07.06.2017)

Rosa, Hartmut (2016a): Beschleunigung. De Veränderung der Zeitstrukturen in der Moderne, 11. Aufl., Frankfurt am Main 2016

Rosa, Hartmut (2016b): Resonanz. Eine Soziologie der Weltbeziehung, Berlin 2016

Rovelli, C. (2018): Die Ordnung der Zeit, Reinbek bei Hamburg 2018

Scheuss, Ralph (2008): Handbuch der Strategien. 220 Konzepte der weltbesten Vordenker, Frankfurt/New York 2008

Schuh, G., Anderl, R., Gausemeier J., ten Hompel, M., Wahlster, W. (2017), (Hrsg.): Industrie 4.0 Maturity Index. Die digitale Transformation von Unternehmen gestalten, (acatech STUDIE), München 2017

Seiwert, L. (2005): Die Bären-Strategie, Kreuzlingen/München 2005

Seiwert, L. (2013): Lass los und du bist Meister deiner Zeit, München 2013

Seiwert, L. (2014): Das 1x1 des Zeitmanagement, 36. Aufl., München 2014

Seiwert, L. (2018): Wenn du es eilig hast, gehe langsam. Mehr Zeit in einer beschleunigten Welt, 17. Aufl., Frankfurt/Main 2018

Simon, Hermann (1989): Die Zeit als strategischer Erfolgsfaktor, in: Zeitschrift für Betriebswirtschaft, 59. Jg., H. 1, 1989, S. 70-93

Stalk, George (1989): Time - next source of competitive advantage, in: McKinsey Quarterly, Spring 1989, pp. 28-50

Stalk, George; Hout, Thomas (1990): Competing Against Time. How Time-Based Competition Is reshaping Global Markets, The Free Press, Collier Macmillan Publishers, New York 1990

Stalk, George; Webber, A. (1993): Japan's Dark Side of Time, in: Harvard Business Review, July-August 1993, pp. 93-102

Suter, Martin (2012): Die Zeit, die Zeit, Diogenes Verlag: Zürich 2012

Taylor, Frederick W. (1919): The principles of scientific management, Harper & Brothers, London 1919

Tersine, R.; Hummingbird, E. (1995): Lead-time reduction: the search for competitive advantage, in: International Journal of Operations & Production Management, Vol. 15, Iss. 2, 1995, pp. 8-18

Time (2018): www.timeanddate.de/kalender/julianischer-gregorianischer, (Retrieved: 28.12.2018)

Time (2019): https://www.timeanddate.de/zeitzonen/schaltsekunden-erklaerung, (Retrieved: 09.03.2019)

Trompenaars, F.; Hampden-Turner, Ch. (1998): Riding the Waves of Culture. Understanding cultural Diversity in Business, 2. Aufl., Nicholas Brealey Publishing Limited, London 1998

Tunc, E.; Gupta, J. (1993): Is Time a Competitive Weapon among Manufacturing Firms?, in: International Journal of Operations & Production Management, Vol. 13, Iss. 3, 1993, pp. 4-12

Virillo, Paul (2015): Rasender Stillstand, Frankfurt am Main 2015

vom Brocke, Jan; Schmiedel, Theresa; Recker, Jan; Trkman, Peter; Mertens, Willem; Viaene, Stijn (2014): Ten principles of good business process management, in: Business Process Management Journal, Vol. 20, Iss. 4, 2014, pp. 530-548

Weber, Jürgen (2012): Logistikkostenrechnung, 3. Aufl., Berlin/Heidelberg 2012

Wegener, Dieter (2017): „Industrie 4.0" – wie die Digitalisierung die Produktionskette revolutioniert, Vortrag auf der 3. INDIGO-Konferenz „Digitale Produktion" an der OTH Amberg-Weiden am 30. Juni 2017

Zugreiseblog (2019): www.zugreiseblog.de/bahn-puenktlichkeit-statistik/ (Retrieved: 07.06.2019)

Zukunftsinstitut (2018a): Zeitalter der Langsamkeit, https://www.zukunftsinstitut.de/artikel/das-zeitalter-der-langsamkeit/, (Retrieved: 06.03.2019)

Zukunftsinstitut (2018b): Langsames Denken, https://www.zukunftsinstitut.de/artikel/slow-thinking-die-kunst-vernetzt-zu-denken/, (Retrieved: 06.03.2019)

Zukunftsinstitut (2018c): Slow TV, https://www.zukunftsinstitut.de/artikel/slow-tv-mehr-zeit-zum-sehen/, (Retrieved: 06.03.2019)

Zukunftsinstitut (2018d): Slow Management, https://www.zukunftsinstitut.de/artikel/slow-management-achtsam-macht-erfolgreich/, (Retrieved: 06.03.2019)